Everyday Mathematics®

The University of Chicago School Mathematics Project

Skills Link

Cumulative Practice Sets
Student Book

Mc Graw Hill **Wright Group**

The **McGraw·Hill** Companies

Photo Credits

Cover—©Tim Flach/Getty Images, cover; Getty Images, cover, *bottom left*
Photo Collage—Herman Adler Design

---

www.WrightGroup.com

 **Wright Group**

Send all inquiries to:
Wright Group/McGraw-Hill
P.O. Box 812960
Chicago, IL 60681

ISBN 978-0-07-622503-3
MHID 0-07-622503-8

4 5 6 7 8 9 MAL 15 14 13 12 11 10

Contents

Practice Sets Correlated to Grade 3 Goals

Content	Everyday Mathematics Grade 3 Grade-Level Goals	Grade 3 Practice Sets
Number and Numeration		
Place value and notation	**Goal 1.** Read and write whole numbers up to 1,000,000; read, write, and model with manipulatives decimals through hundredths; identify places in such numbers and the values of the digits in those places; translate between whole numbers and decimals represented in words, in base-10 notation, and with manipulatives.	1, 3, 7, 15, 33, 35, 37, 39, 40, 41, 42, 45, 51, 55, 59, 63, 77, 79, 81, 89
Meanings and uses of fractions	**Goal 2.** Read, write, and model fractions; solve problems involving fractional parts of a region or a collection; describe strategies used.	64, 66, 67, 70, 71, 73, 74, 76, 83, 84, 90, 97
Number theory	**Goal 3.** Find multiples of 2, 5, and 10.	33, 59
Equivalent names for whole numbers	**Goal 4.** Use numerical expressions involving one or more of the basic four arithmetic operations to give equivalent names for whole numbers.	4, 11, 59, 70
Equivalent names for fractions, decimals, and percents	**Goal 5.** Use manipulatives and drawings to find and represent equivalent names for fractions; use manipulatives to generate equivalent fractions.	66, 69, 72, 73, 76, 78
Comparing and ordering numbers	**Goal 6.** Compare and order whole numbers up to 1,000,000; use manipulatives to order decimals through hundredths; use area models and benchmark fractions to compare and order fractions.	1, 7, 38, 45, 68, 82
Operations and Computation		
Addition and subtraction facts	**Goal 1.** Demonstrate automaticity with all addition and subtraction facts through 10 + 10; use basic facts to compute fact extensions such as 80 + 70.	1, 4, 6, 10, 11, 13, 14, 15, 17, 19, 22, 28, 32, 42, 48, 54
Addition and subtraction procedures	**Goal 2.** Use manipulatives, mental arithmetic, paper-and-pencil algorithms, and calculators to solve problems involving the addition and subtraction of whole numbers and decimals in a money context; describe the strategies used and explain how they work.	5, 6, 7, 9, 13, 14, 15, 16, 17, 18, 22, 25, 26, 28, 37, 38, 41, 42, 48, 54, 56, 61, 62, 65, 82, 97
Multiplication and division facts	**Goal 3.** Demonstrate automaticity with $\times 0$, $\times 1$, $\times 2$, $\times 5$, and $\times 10$ multiplication facts; use strategies to compute remaining facts up to 10×10.	25, 30, 31, 32, 33, 35, 38, 43, 44, 48, 49, 54, 56, 57, 58, 60, 63, 69, 72, 75, 76, 81, 84, 87, 89, 91, 96
Multiplication and division procedures	**Goal 4.** Use arrays, mental arithmetic, paper-and-pencil algorithms, and calculators to solve problems involving the multiplication of 2- and 3-digit whole numbers by 1-digit whole numbers; describe the strategies used.	32, 73, 75, 76, 77, 80, 81, 82
Computational estimation	**Goal 5.** Make reasonable estimates for whole number addition and subtraction problems; explain how the estimates were obtained.	9, 14, 16, 17, 44
Models for the operations	**Goal 6.** Recognize and describe change, comparison, and parts-and-total situations; use repeated addition, arrays, and skip counting to model multiplication; use equal sharing and equal grouping to model division.	13, 14, 26, 27, 28, 29, 42, 47, 52, 55, 56, 58, 59, 60, 76, 77, 79

Content	*Everyday Mathematics* Grade 3 Grade-Level Goals	Grade 3 Practice Sets
Data and Chance		
Data collection and representation	**Goal 1.** Collect and organize data or use given data to create charts, tables, bar graphs, and line plots.	46, 92
Data analysis	**Goal 2.** Use graphs to ask and answer simple questions and draw conclusions; find the maximum, minimum, range, mode, and median of a data set.	3, 46, 62, 88, 89, 90, 92
Qualitative probability	**Goal 3.** Describe events using *certain, very likely, likely, unlikely, very unlikely, impossible* and other basic probability terms; explain the choice of language.	65, 95
Quantitative probability	**Goal 4.** Predict the outcomes of simple experiments and test the predictions using manipulatives; express the probability of an event by using "__out of __" language.	94, 95, 96, 97
Measurement and Reference Frames		
Length, weight, and angles	**Goal 1.** Estimate length with and without tools; measure length to the nearest $\frac{1}{2}$ inch and $\frac{1}{2}$ centimeter; draw and describe angles as records of rotations.	2, 19, 20, 34, 48, 56, 60, 67, 74, 84
Area, perimeter, volume, and capacity	**Goal 2.** Describe and use strategies to measure the perimeter of polygons; count unit squares to find the areas of rectangles.	21, 22, 23, 24, 25, 27, 49, 51, 60
Units and systems of measurement	**Goal 3.** Describe relationships among inches, feet, and yards; describe relationships between minutes in an hour, hours in a day, days in a week.	20, 36, 49, 84, 85, 86, 88
Time	**Goal 4.** Tell and show time to the nearest minute on an analog clock; tell and write time in digital notation.	2, 9, 21, 49, 50
Geometry		
Lines and angles	**Goal 1.** Identify and draw points, intersecting and parallel line segments and lines, rays, and right angles.	47, 48, 51, 52, 62
Plane and solid figures	**Goal 2.** Identify, describe, model, and compare plane and solid figures including circles, polygons, spheres, cylinders, rectangular prisms, pyramids, cones, and cubes using appropriate geometric terms including the terms *face, edge, vertex,* and *base*.	2, 21, 25, 40, 50, 54, 55, 81, 86
Transformations and symmetry	**Goal 3.** Create and complete two-dimensional symmetric shapes or designs; locate multiple lines of symmetry in a two-dimensional shape.	53, 60
Patterns, Functions, and Algebra		
Patterns and functions	**Goal 1.** Extend, describe, and create numeric patterns; describe rules for patterns and use them to solve problems; use words and symbols to describe and write rules for functions involving addition, subtraction, and multiplication and use those rules to solve problems.	6, 8, 10, 12, 29, 35, 36, 44, 55, 57, 63, 68, 72, 73, 87, 95
Algebraic notation and solving number sentences	**Goal 2.** Read, write, and explain number sentences using the symbols $+$, $-$, $\times$, $\div$, $=$, $>$, and $<$; solve number sentences; write expressions and number sentences to model number stories.	6, 39, 61, 69, 71, 75, 80, 90, 94
Order of operations	**Goal 3.** Recognize that numeric expressions can have different values depending on the order in which operations are carried out; understand that grouping symbols can be used to affect the order in which operations are carried out.	59, 69
Properties of arithmetic operations	**Goal 4.** Describe and apply the Commutative and Associative Properties of Addition, the Commutative Property of Multiplication, and the Multiplicative Identity.	59, 67

Grade 2 Review: Number and Numeration

Count on.

1. 800; 900; _____; _____; _____; _____

2. 675; 700; _____; _____; _____; _____

Count back.

3. 624; 623; _____; _____; _____; _____

4. 401; 400; _____; _____; _____; _____

5. Write the number.

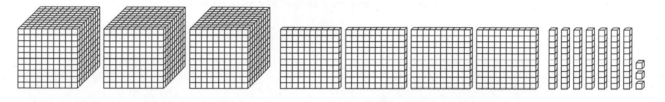

_____

Look at the number 5,483.

6. In what place is the 8? _____

7. In what place is the 5? _____

8. In what place is the 3? _____

9. In what place is the 4? _____

10. How much money? Write the amount.

$_____ . _____

Grade 2 Review: Number and Numeration

11. Circle $\frac{1}{3}$ of each group.

12. Shade 3 parts of each figure. Write a fraction for the shaded part.

_____ _____ _____ _____

13. Write the fractions $\frac{3}{6}$, $\frac{3}{8}$, $\frac{3}{3}$, and $\frac{3}{5}$ in order from least to greatest.

_____

Circle *odd* or *even* for each number.

14. 521   odd   even   **15.** 732   odd   even

16. 800   odd   even   **17.** 609   odd   even

18. Show 7 two different ways.   **19.** Show 12 two different ways.

20. Shade $\frac{1}{2}$ of each figure. Write the fraction below.

_____ _____ _____ _____

Write in order from least to greatest.

21. 430   5,207   61   897   9   **22.** 683   638   623   268   326

_____   _____

_____   _____

Grade 2 Review: Operations and Computation

Find the sum or difference.

1. $15 + 3 =$ _____

2. $8 + 8 =$ _____

3. $0 + 13 =$ _____

4. $2 + 17 =$ _____

5. $6 + 6 =$ _____

6. $19 + 0 =$ _____

7. $18 - 5 =$ _____

8. $9 - 9 =$ _____

9. $14 - 0 =$ _____

Solve.

10. Ms. Cathay's class of 28 students and Mr. Patel's class of 31 students went to the zoo. About how many students went to the zoo altogether? Explain how you made your estimate.

_____

_____

11. Tia gave the cashier 3 quarters, 4 dimes, 5 nickels, and 7 pennies to pay for her stickers. Ken paid $1.58 for his stickers. Whose stickers cost more? _____

How much more did the stickers cost? _____

12. There are 8 crayons in each box. There are 3 boxes. Draw a picture to show how many crayons there are altogether.

Write a number sentence for your picture.

_____

13. There are 21 beads. Three children want to use the beads to make necklaces. Draw a picture to show how many beads each child can have if the beads are divided equally.

Write a number sentence for your picture.

_____

Grade 2 Review: Data and Chance

A third grade class had a magazine sale. Megan sold 4 magazines.
Hoyt sold 9. Graham sold 4 fewer magazines than Hoyt. Bisa sold
two more magazines than Megan. Sassandra sold 7 magazines.

1. Make a tally chart of the number of
magazines each child sold.

Number of Magazines Sold	
Megan	
Hoyt	
Graham	
Bisa	
Sassandra	

2. Use the data from the tally chart
to complete the bar graph.

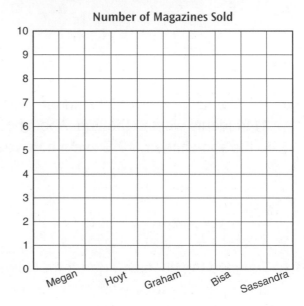

Number of Magazines Sold

Use the bar graph to answer the questions.

3. What is the maximum number of magazines sold? _____ magazines

4. What is the minimum number of magazines sold? _____ magazines

5. What is the median of the data set? _____ magazines

Look at the spinner. Write *certain, likely, unlikely,* or *impossible* for each event.

6. Spinning a 1, 2, or 3 _____

7. Spinning a 1 _____

8. Spinning a 3 _____

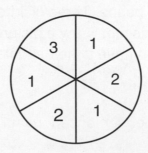

9. Spinning a 4 _____

Grade 2 Review: Measurement and Reference Frames

Count the squares to find the area.

1.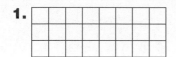

2.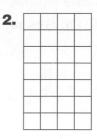

_____ square units   _____ square units

3. This line segment is 2 centimeters long.
Draw a line segment that is about 3 times longer.

About how long is it? About _____ centimeters

4. About how long is the pencil? About _____ inches

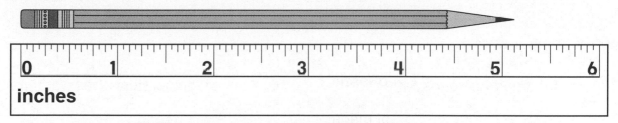

5. Tyrone wants to go to basketball camp. Camp lasts 4 days.
If camp begins on Sunday, on what day will it end?

_____

6. Cheri spends 6 hours at school, 1 hour working on homework,
1 hour practicing the violin, 1 hour playing soccer, 1 hour
reading, 1 hour helping with the chores around the house,
and 10 hours sleeping. How many hours are left in Cheri's day?

_____

7. Show how to make $1.49, using the **fewest** number of coins.
Use Ⓟ, Ⓝ, Ⓓ, and Ⓠ.

8. Show another way to make $1.49.

Grade 2 Review: Measurement and Reference Frames

Draw the hour hand and the minute hand.

9.

2:10

10.

6:55

11.

12:40

Write the time.

12.

_____

13.

_____

14.

_____

Read the temperatures.

15. At what temperature does water boil?

Water Boils

Celsius
120
110
100
90
80
70
60

Fahrenheit
230
220
210
200
190
180
170

_____ °C   _____ °F

16. At what temperature does water freeze?

Water Freezes

Celsius
50
40
30
20
10
0
−10

Fahrenheit
80
70
60
50
40
30
20

_____ °C   _____ °F

Grade 2 Review: Geometry

1. Circle the parallel lines.

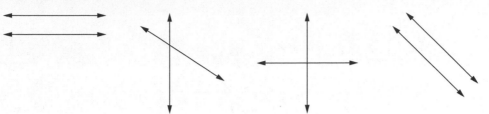

Draw the figure.

    2. circle   **3.** square   **4.** triangle   **5.** rectangle

Name the figure.

6. 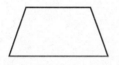   **7.**   **8.**

_____   _____   _____

   9. Circle the sphere.   **10.** Circle the cone.   **11.** Circle the cube.

   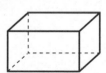

Draw the other half of each shape.

12. 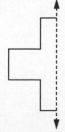   **13.**

Name	Date	Time

Grade 2 Review: Patterns, Functions, and Algebra

Find the rule. Complete the table.

1.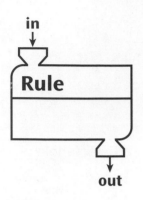

in	out
	4
7	14
9	
10	20
	40

2.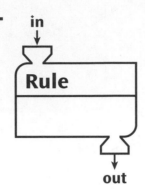

in	out
5	2
	7
12	
15	
20	17

3. Draw the next three shapes in the pattern.

○ ☐ ☐ △ ○ ☐ _____ _____ _____

Write <, >, or =.

4. 18 + 15 _____ 15 + 18

5. 20 − 12 _____ 5 + 4

6. (2 + 9) + 7 _____ 2 + (9 + 7)

7. 1,235 _____ 1,325

8. 389¢ _____ $38.90

9. $2.53 _____ 1 dollar, 5 quarters, 1 dime, 2 nickels, and 3 pennies

10. Jamal has 3 quarters, 5 dimes, 3 nickels, and 8 pennies. He wants to buy a notebook for $1.29. Use Ⓠ, Ⓓ, Ⓝ, and Ⓟ to show how much money he has.

Cross out how much Jamal will spend on the notebook.

How much money does he have left? _____

Write the turn-around facts.

11. 16 + 8 = _____ + _____

12. 4 + 17 = _____ + _____

13. 5 + 9 = _____ + _____

14. 25 + 2 = _____ + _____

Practice Set 1

Write the missing number.

1.
	598	
607		609

2.
24		26
44		46

3.
179
199

4.
	707
	717
	727

5.
444	
	466

6.
998	999	

7. Put these numbers in order from smallest to largest:

259   262   260   258   263   261

_____

8. Put these numbers in order from largest to smallest:

990   980   1,000   1,100   970   1,200

_____

Count by 2s. Find the missing numbers.

9. 31, 33, _____ , _____, _____, 41, _____, _____, _____, 49

10. 92, _____, 96, _____, 100, _____, _____, 106, _____, _____

11. 131, 133, _____, _____, 139, _____, _____, 145, _____, _____

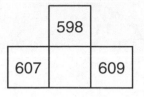 **Add. Remember to practice and memorize your addition facts.**

12. $2 + 4 =$ _____

13. $5 + 3 =$ _____

14. $9 + 1 =$ _____

15. $6 + 7 =$ _____

16. $2 + 8 =$ _____

17. $3 + 6 =$ _____

Practice Set 2

Record the time shown on each clock.

1.

_____

2.

_____

3.

_____

4.

_____

5.

_____

6.

_____

7.

_____

8.

_____

9.

_____

Use with or after Lesson 1•4.

Practice Set 2 *continued*

SRB
102–109

Measure the line segment above in inches and centimeters.

10. The line segment is _____ inches long.

11. The line segment is _____ centimeters long.

Write the name of each shape.

12.

_____

13.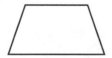

_____

14.

_____

15.

_____

16.

_____

17.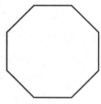

_____

18.

_____

19.

_____

20.

_____

Practice Set 3

Use the tally chart to answer each question.

Ice Cream Favorites	
Kind of Ice Cream	**Number of Students**
Vanilla	~~HHH~~
Strawberry	~~HHH~~ //
Double Chocolate	~~HHH~~ ~~HHH~~ //
Chocolate Chip Mint	////
Maple Nut	///

1. How many students like Double Chocolate ice cream best? _____

2. What is the favorite flavor? _____

3. What is the least favorite flavor? _____

4. How many students altogether chose Double Chocolate or

   Chocolate Chip Mint? _____

5.  **Writing/Reasoning** Use the tally chart to write your own
   question. Then write the answer.

   _____

   _____

Write the number shown by the base-10 blocks.

6.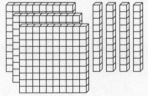

7.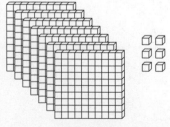

8.

_____   _____   _____

Practice Set **3** continued

Use the bar graph to answer each question.

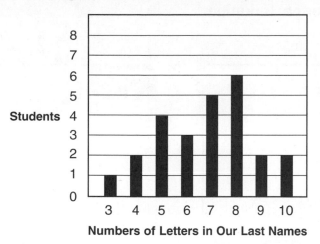

Numbers of Letters in Our Last Names

9. How many students have eight letters in their last name? _____

10. How many students have four letters in their last name? _____

11. How many students' names are represented in the bar graph? _____

12. How many students have fewer than six letters in their last name? _____

13. How many students have more than six letters in their last name? _____

14. How many letters does the *longest* name have? (This is called the *maximum*.)

   _____

15. How many letters does the *shortest* name have? (This is called the *minimum*.)

   _____

16. What is the *range* of the numbers of letters? _____

17. What is the *mode* of this set of data? _____

(*Hint:* If you don't remember what range and mode are, look them up in your *Student Reference Book.*)

FACTS PRACTICE **Add or subtract. Remember to practice and memorize your basic facts.**

18. 4 + 5 = _____ 19. 8 + 10 = _____ 20. 2 + 9 = _____

21. 12 − 9 = _____ 22. 9 − 3 = _____ 23. 15 − 2 = _____

Practice Set 4

Make your own name-collection box for each of the five numbers listed below. Include 10 different names for each number.

Example

```
┌─────────────────┐
│ 16              │
│                 │
│ 1 ten 6 ones  8 │
│             + 8 │
│ 16 ÷ 1          │
│                 │
│ sixteen   5 + 4 + 7 │
│                 │
│ 7 + 9   ЖЖ ЖЖ ЖЖ I │
│                 │
│ 4 × 4    10 + 6 │
│ • • • • • • • • │
│ • • • • • • • • │
└─────────────────┘
```

1.
```
┌─────────────────┐
│ 9               │
│                 │
│                 │
│                 │
│                 │
│                 │
│                 │
└─────────────────┘
```

2.
```
┌─────────────────┐
│ 14              │
│                 │
│                 │
│                 │
│                 │
│                 │
│                 │
└─────────────────┘
```

3.
```
┌─────────────────┐
│ 8               │
│                 │
│                 │
│                 │
│                 │
│                 │
│                 │
└─────────────────┘
```

4.
```
┌─────────────────┐
│ 12              │
│                 │
│                 │
│                 │
│                 │
│                 │
│                 │
└─────────────────┘
```

5.
```
┌─────────────────┐
│ 10              │
│                 │
│                 │
│                 │
│                 │
│                 │
│                 │
└─────────────────┘
```

6. Count by 3s.

3, 6, 9, 12, _____, _____, _____, _____, _____, _____

7. Count back by 10s.

140, 130, 120, _____, _____, _____, _____, _____, _____

 Add. Remember to practice and memorize your addition facts.

8. $7 + 8 =$ _____

9. $6 + 9 =$ _____

10. $9 + 5 =$ _____

11. $12 + 8 =$ _____

12. $6 + 8 =$ _____

13. $8 + 9 =$ _____

Practice Set 5

Use the number grid to answer the problems below.

									0
1	2	3	4	5	6	7	8	9	10
11	12	13	14	15	16	17	18	19	20
21	22	23	24	25	26	27	28	29	30
31	32	33	34	35	36	37	38	39	40
41	42	43	44	45	46	47	48	49	50
51	52	53	54	55	56	57	58	59	60
61	62	63	64	65	66	67	68	69	70
71	72	73	74	75	76	77	78	79	80
81	82	83	84	85	86	87	88	89	90
91	92	93	94	95	96	97	98	99	100
101	102	103	104	105	106	107	108	109	110

1. Find 20 more than 84. _____

2. Find 16 more than 68. _____

3. Find 12 less than 32. _____

4. Find 35 less than 44. _____

5. Start at 0 and count by 3s along the *second* row of the number grid.
Write the numbers from your count.

_____

6. Start at 41 and count by 3s along the *sixth* row of the number grid.
Write the numbers from your count.

_____

7. Start at 81 and count by 6s along two rows of the number grid.
Write the numbers from your count.

_____

Practice Set 6

Use your calculator to count by 10s. Find the missing numbers.

> **Example**   40, _____, _____, 70, 80, _____, _____
>
> **Press:**
>
> **Display:** 50, 60, 70, 80, 90, 100

1. 25, _____, 45, _____, _____, _____, 85, _____, _____, _____

2. 123, _____, 143, _____, _____, _____, _____, _____, _____, 213

Use your calculator to solve each problem.

3. The first permanent English colony was established in the New World in 1607. The colonies united to demand freedom from England in 1776. How many years went by before the colonies demanded freedom?

_____

4. Marta read a book that was 45 pages long. Next, she read a book that was 82 pages long. Then she read a book that was 106 pages long. How many pages did Marta read in all?

_____

Write the missing numbers.

5. 10 = _____ + 4   **6.** 5 + _____ = 10   **7.** _____ + 7 = 10

8. 26 + _____ = 30   **9.** 50 = _____ + 43   **10.** 81 + _____ = 90

11. **Writing/Reasoning** Christine was born in 1966. Her grandmother was born in 1900. How old was her grandmother when Christine was born? Write a number model and explain your answer.

_____

_____

Practice Set 7

Estimate to answer _yes_ or _no_.

1. You have $5.00. Do you have enough to buy a notebook for $3.99 and a pen for $1.55?

_____

2. You have $4.50. Do you have enough to buy two boxes of pencils that cost $2.10 each?

_____

3. You have $10.00. Do you have enough to buy crayons for $1.89, a backpack for $6.98, and paper clips for 79¢?

_____

4. You have $3.20. Do you have enough to buy a marker for $1.79 and a pad of paper for $1.49?

_____

5. **Writing/Reasoning** Explain how you found your estimate in Problem 4.

_____

_____

Solve each problem.

6. Juana paid for a video that cost $6.59 with a $10.00 bill. How much change did she receive?

_____

7. Larry's lunch cost $3.25. Larry paid for his lunch with a $5.00 bill. How much change did he receive?

_____

Write the letter that identifies each amount.

8. $\frac{1}{2}$ dime _____ **A.** dime

9. quarter _____ **B.** $0.50

10. $\frac{1}{10}$ dollar _____ **C.** $\frac{3}{4}$ dollar

11. $0.01 _____ **D.** penny

12. $\frac{1}{2}$ dollar _____ **E.** nickel

13. $0.75 _____ **F.** $\frac{1}{4}$ dollar

Practice Set 7 continued

Write <, >, or =.

14. $1.59 _____ $0.95

15. $7.52 _____ $4.75

16. $0.88 _____ $1.08

17. $6.65 _____ $5.66

| = means *is equal to* |
| < means *is less than* |
| > means *is greater than* |

18. $10.01 _____ $9.10   **19.** $0.75 _____ 75 cents

20. $1.11 _____ 111 pennies   **21.** 63 cents _____ $1.63

22. $4.84 _____ $4.48   **23.** 5 nickels _____ $0.20

Enter each amount of money on your calculator. Then write the equivalent value you see on your calculator display.

Example   Enter: 93¢   Display shows: 0.93

24. $0.08 _____   **25.** $1.59 _____   **26.** 98¢ _____

27. $6.57 _____   **28.** 3¢ _____   **29.** 59¢ _____

30. $2.43 _____   **31.** $0.79 _____   **32.** $3.04 _____

Draw coins to show each amount of money in two different ways.

Example   87¢

| Q Q Q | D D D D D D |
| D P P | N N N N N P P |

33. 42¢   **34.** $0.35

35. 27¢   **36.** $0.54

Practice Set 8

Complete each Frames-and-Arrows diagram.

Example

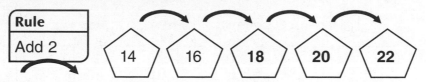

Rule
Add 2

14 16 **18** **20** 22

1.

Rule
+10

10 20 ☐ ☐ ☐ ☐

2.

Rule
−2

28 26 ⬡ ⬡ ⬡

3.

Rule
Subtract 10

90 80 ◯ ◯ ◯

Find each missing number. There may be more than one correct answer.

4. 1 Ⓠ = _____ ¢

5. 20 Ⓟ = _____ ¢

6. _____ Ⓓ = $0.60

7. _____ Ⓠ = $1

8. 1 half-dollar = _____ ¢

9. _____ Ⓠ = 10 Ⓝ

10. 1 quarter = _____ dime(s) and _____ nickel(s)

11. _____ dime(s) and _____ pennies = 1 dollar

12. 82¢ = _____ quarter(s), _____ nickel(s), and _____ pennies

Practice Set 9

SRB
57
200 201

Complete each Frames-and-Arrows diagram.

1.

Rule
−4

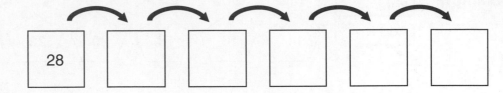

28 ☐ ☐ ☐ ☐ ☐

2.

Rule
+8

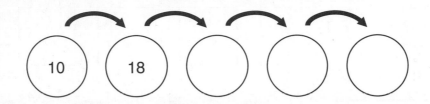

10 18 ○ ○ ○

Solve.

3. Katie started reading her book at 3:30 P.M. She finished reading at 5:10 P.M. How long did she read? _____

4. A new movie, *The Wright Brothers,* begins at the time shown on the first clock and ends at the time on the second clock. How long is the movie? _____

Make a ballpark estimate. Write a number model for your estimate. Then find the exact answer.

5. Ballpark estimate: _____

$$124 + 380$$

6. Ballpark estimate: _____

$$287 + 111$$

7. Ballpark estimate: _____

$$238 - 102$$

8. Ballpark estimate: _____

$$341 - 120$$

Use with or after Lesson 1·12.

Practice Set 10

Write the number family for each Fact Triangle.

Example

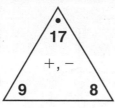

$$9 + 8 = 17$$
$$8 + 9 = 17$$
$$17 - 9 = 8$$
$$17 - 8 = 9$$

1.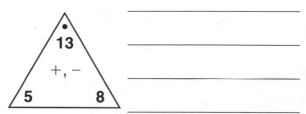

_____
_____
_____
_____

2.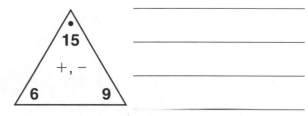

_____
_____
_____
_____

3.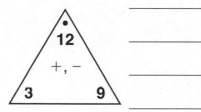

_____
_____
_____
_____

4.

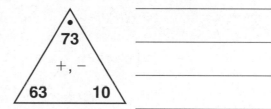

_____
_____
_____
_____

Write the addition and subtraction facts for each group of numbers.

Example 7, 14, 7 $7 + 7 = 14$
                         $14 - 7 = 7$

5. 8, 8, 16

_____

_____

6. 9, 9, 18

_____

_____

7. 20, 20, 40

_____

_____

8. 28, 14, 14

_____

_____

Practice Set 10 continued

SRB
200 201

Use two rules for each set of Frames and Arrows.
Write the numbers for the empty frames.

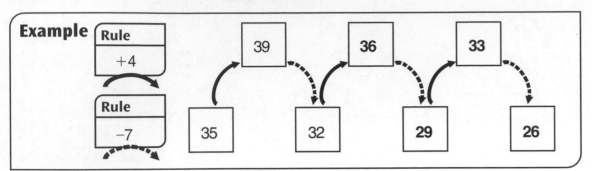

Example

Rule +4

Rule −7

39   36   33

35   32   29   26

9.

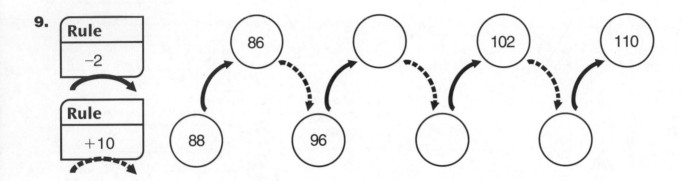

Rule −2

Rule +10

86   102   110

88   96

10.

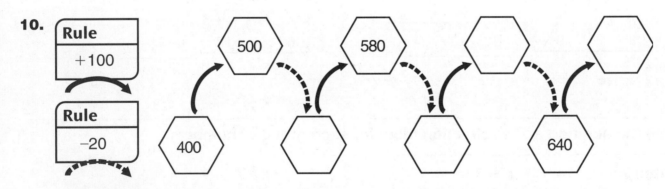

Rule +100

Rule −20

500   580

400   640

11.

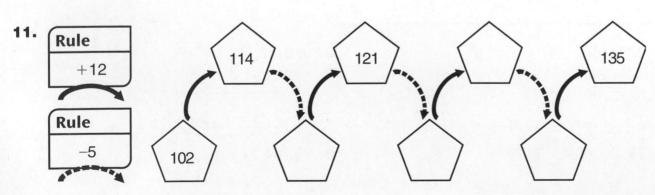

Rule +12

Rule −5

114   121   135

102

Use with or after Lesson 2·1.

Practice Set 11

SRB
254 255

Find each missing number.

1. 6 + 8 = _____

60 + 80 = _____

600 + 800 = _____

2. _____ = 12 − 5

_____ = 120 − 50

_____ = 1,200 − 500

3. 9 + _____ = 13

90 + _____ = 130

900 + _____ = 1,300

4. 4 = 9 − _____

40 = 90 − _____

400 = 900 − _____

5. _____ − 8 = 7

_____ − 80 = 70

_____ − 800 = 700

6. 8 = _____ − 9

80 = _____ − 90

800 = _____ − 900

Use addition or subtraction to complete each problem on your calculator. Tell how much you added or subtracted.

Example	Enter	Change to	What I did
	34	50	+ 16

	Enter	Change to	What I did
7.	90	72	_____
8.	22	50	_____
9.	100	58	_____
10.	200	120	_____
11.	130	250	_____
12.	900	400	_____

Use with or after Lesson 2·2.

Practice Set **11** *continued*

Circle the names that DO NOT belong in each name-collection box.

Example

15

(5 less than 19)

5×3

(4 × 4)   (8 + 6)

$3 + 3 + 3 + 3 + 3$

(7 + 6 + 3)   𝈫𝈫𝈫 (tally marks)

$5 + 5 + 5$

2 more than 13

13.

16

six	4	9
	× 4	× 2

16×1

$10 + 6$ 3×5

$6 + 6 + 5$

2 more than 14

3 less than 20

14.

14

$15 - 1$	7
$14 + 1$	+ 7

$0 + 14$ •••••••
                   •••••••

7×7

$5 + 4 + 6$ 𝈫𝈫 ////

$9 + 6$ $10 + 4$

15.

12

$12 + 1$	3
$20 - 8$	× 4

$3 + 10$ 1×12

$5 + 8$ 2×6

•••
•••   𝈫 ///
•••

16.

20

twenty 4×5

𝈫𝈫𝈫 ////

$13 + 8$

$20 + 0$

$5 + 6 + 9$ $10 + 10$

20×0

$8 + 7 + 7$

Practice Set 12

Find each missing number.

Example

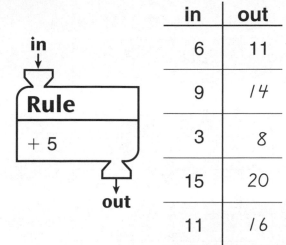

in	out
6	11
9	14
3	8
15	20
11	16

1.

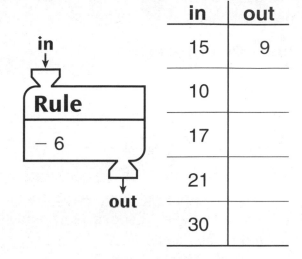

in	out
15	9
10	
17	
21	
30	

2.

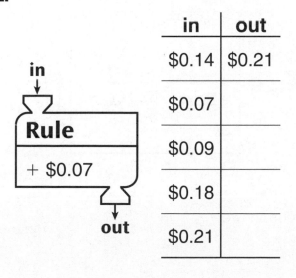

in	out
$0.14	$0.21
$0.07	
$0.09	
$0.18	
$0.21	

3.

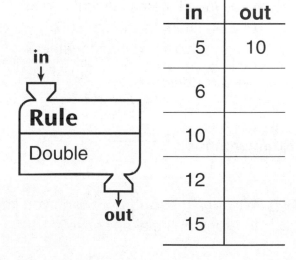

in	out
5	10
6	
10	
12	
15	

Practice Set 13

SRB
256 257

Use the parts-and-total diagram to help you solve the problem.

1. The bakery shop sold 36 cupcakes. The next day the shop sold 48 cupcakes. How many cupcakes were sold in all?

   Answer the question:

   _____

   Number model:

   _____

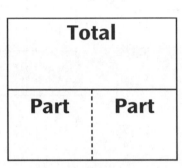

Total	
Part	Part

2. Sarah and Nick are shelving books at the school library. Nick shelves 65 books and Sarah shelves 83 books. How many books do they shelve in all?

   Answer the question:

   _____

   Number model:

   _____

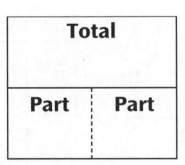

Total	
Part	Part

3. Lauren buys one item that costs $0.89 and a second item that costs $0.75. How much do the two items cost together?

   Answer the question:

   _____

   Number model:

   _____

Total	
Part	Part

4. **Writing/Reasoning** How do you know your answer to Problem 3 makes sense?

   _____

   _____

FACTS PRACTICE   **Add or subtract. Remember to practice and memorize your basic facts.**

5. $8 + 9 =$ _____

6. $9 - 1 =$ _____

7. $13 - 5 =$ _____

8. $10 + 5 =$ _____

9. $11 - 2 =$ _____

10. $12 + 4 =$ _____

Practice Set 14

Use the change diagram to solve the problem.
Write the answer and a number model.

1. Jeremy saves $17. He receives a gift
card for $25. How much does he have
to spend in all?

Answer the question:

_____

Number model:

_____

2. Maria bought a pair of jeans for $17.50.
She gave the clerk a $20 bill. How much
change did she receive?

Answer the question:

_____

Number model:

_____

3. **Writing/Reasoning** Can you write a different number model
for Problem 2? Explain your answer.

_____

_____

Make a ballpark estimate. Write a number model to show
your estimate. Then find the exact answer.

4. Ballpark estimate: _____

     84
   + 90

5. Ballpark estimate: _____

     68
   − 37

6. Ballpark estimate: _____

     38
   − 11

7. Ballpark estimate: _____

     57
   + 68

Practice Set 15

Choose one of the diagrams to help you solve each number story.

1. Bedelia picked 23 flowers on Tuesday. She picked more flowers on Wednesday. She picked a total of 47 flowers. How many flowers did she pick Wednesday?

Answer the question:

_____

Number model:

_____

2. Sara spelled 83 words correctly at this year's spelling contest. Last year she spelled only 47 words correctly. What is the difference between her two scores?

Answer the question:

_____

Number model:

_____

3. Larissa made 67 paper birds for a craft fair. Then she made 38 paper insects. How many objects did she make in all?

Answer the question:

_____

Number model:

_____

Quantity

Quantity
_____
Difference

Total	
Part	**Part**

Change

Start	→	End
	_____	

32

Practice Set **15** *continued*

Count by 10s. Find the missing numbers.

4. 400, 410, _____, 430, _____, _____, 460, _____, _____, _____, 500

5. 1,010; 1,020; _____; 1,040; _____; _____; 1,070; _____; _____; 1,100

6. 3,225; 3,235; _____; _____; 3,265; _____; _____; _____; 3,305; _____

7. 8,712; 8,722; _____; 8,742; _____; _____; 8,772; _____; _____; 8,802

8. 3,218; 3,228; _____; _____; 3,258; _____; _____; _____; 3,298; _____

Tell what the underlined digit stands for in each number.

Example	9,<u>6</u>13	**600**

9. <u>2</u>,917 _____ **10.** 3,04<u>6</u> _____ **11.** 8<u>5</u>1 _____

12. 8,<u>0</u>46 _____ **13.** <u>5</u>,425 _____ **14.** <u>1</u>4,523 _____

15. 6,79<u>1</u> _____ **16.** 4,3<u>8</u>0 _____ **17.** 6<u>3</u>,941 _____

Write one addition fact and one subtraction fact for each group of numbers.

Example 5, 17, 22	**5 + 17 = 22; 17 + 5 = 22** **22 − 5 = 17; 22 − 17 = 5**

18. 28, 9, 37 _____ **19.** 50, 30, 80 _____

_____ _____

20. 6, 57, 63 _____ **21.** 60, 8, 52 _____

_____ _____

22. 70, 90, 160 _____ **23.** 400, 500, 900 _____

_____ _____

Practice Set 16

Estimate first. Then use the partial-sums addition method to add.

Example

Ballpark estimate:

$470 + 20 = 490$

```
     100s │ 10s │ 1s
       4  │  6  │  7
    +     │  1  │  8
    ──────┼─────┼────
       4  │  0  │  0
          │  7  │  0
          │  1  │  5
    ──────┼─────┼────
       4  │  8  │  5
```

1. Ballpark estimate:

_____

```
  345
+  69
```

2. Ballpark estimate:

_____

```
  38
+ 45
```

3. Ballpark estimate:

_____

```
   75
+ 129
```

Name   Date   Time

Practice Set 17

Estimate first. Then use the trade-first subtraction method to subtract.

Example

Ballpark estimate:

225 – 175 = 50

100s	10s	1s
/	/2	
2̷	2̷	5
– 1	7	3
	5	2

1. Ballpark estimate:

  305
– 69

2. Ballpark estimate:

  138
– 45

3. Ballpark estimate:

  275
– 129

Practice Set **17** *continued*

Find each missing number.

4. $\boxed{\$1}$ = _____ Ⓓ

5. $0.42 = _____ Ⓟ

6. _____ half-dollars = $2.00

7. _____ Ⓝ = 4 Ⓓ

8. 75¢ = _____ Ⓠ

9. _____ Ⓓ = 1 half-dollar

10. _____ Ⓝ = 35¢

11. $0.70 = _____ Ⓓ

COMPUTATION PRACTICE **Add or subtract.**

12. 52 + 79 = _____

13. 98 − 46 = _____

14. 14 + 24 + 36 = _____

15. 81 − 49 = _____

16. 26 + 91 = _____

17. 65 − 28 = _____

18. 104
        − 67

19.   26
        + 13

20.   79
        − 24

21.   57
        + 26

Count by 100s. Find the missing numbers.

22. 1,000; 1,100; _____; 1,300; _____; _____; 1,600; _____; _____; 1,900

23. 2,450; 2,550; _____; _____; 2,850; _____; 3,050; _____; _____; 3,350

24. 7,304; 7,404; _____; _____; _____; 7,804; _____; 8,004; _____; _____

25. 5,416; _____; _____; 5,716; _____; _____; _____; 6,116; 6,216; 6,316

26. 2,883; _____; _____; 3,183; 3,283; _____; _____; _____; 3,683; _____

Use with or after Lesson 2•8.

Practice Set 18

Use the diagrams below to help you solve the problems.

Total		
Part	**Part**	**Part**

Total			
Part	**Part**	**Part**	**Part**

1. Samuel bought presents for 40 cents, 50 cents, 60 cents, and 70 cents. How much money did he spend in all?

   _____

   Check: Does my answer make sense? _____

2. Trina rode her bicycle 12 miles Friday. She rode 14 miles Saturday and 15 miles Sunday. How many miles did she ride in all?

   _____

   Check: Does my answer make sense? _____

3. Jon, Dave, and Kevin collected rocks at the beach. Each boy collected 25 rocks. How many rocks did the boys collect in all?

   _____

   Check: Does my answer make sense? _____

4. The Torrey family was on vacation. One day, they spent $140 for a motel room, $130 for meals, and $200 at a park. How much money did they spend that day?

   _____

   Check: Does my answer make sense? _____

Practice Set 18 continued

Find the missing numbers for each addition and subtraction puzzle.

Example

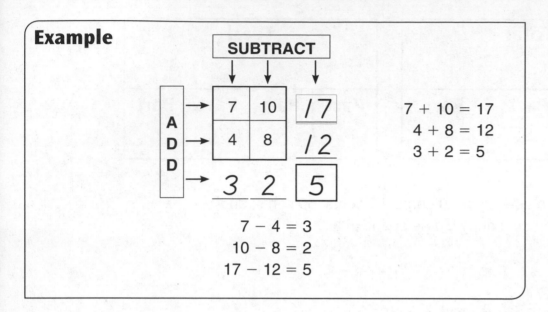

$7 + 10 = 17$
$4 + 8 = 12$
$3 + 2 = 5$

$7 - 4 = 3$
$10 - 8 = 2$
$17 - 12 = 5$

5.

SUBTRACT

35	29	
18	15	

6.

SUBTRACT

53	34	
39	21	

7.

SUBTRACT

	75	
		15
25	70	

8.

SUBTRACT

84		137
64		
		56

Practice Set 19

Measure each line segment to the nearest $\frac{1}{4}$ inch.

1. ├──────────────────┤

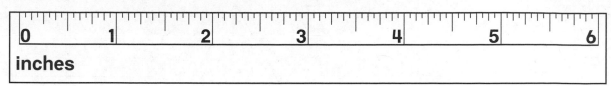

_____

2. ├──────────────────────────────┤

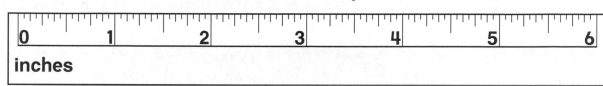

_____

3. ├────────────────────────┤

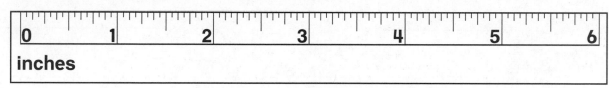

_____

4. ├──────────────────┤

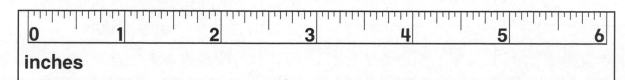

_____

5. ├──────────────┤

_____

Practice Set 19 continued

SRB
54 55

Find the missing number for each Fact Triangle. Then write the
family of facts for that triangle.

6.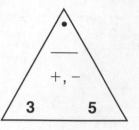

_____

_____

_____

7.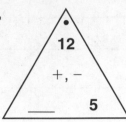

_____

_____

_____

8.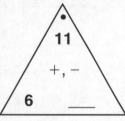

_____

_____

_____

9.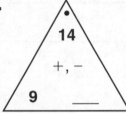

_____

_____

_____

10.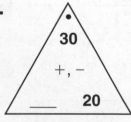

_____

_____

_____

11.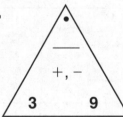

_____

_____

_____

12.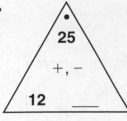

_____

_____

_____

13.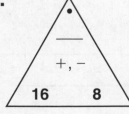

_____

_____

_____

Use with or after Lesson 3•2.

Practice Set 20

Use measuring tools if you need help answering these questions:

1. How many inches are in 1 foot? _____

2. How many inches are in 1 yard? _____

3. How many feet are in 1 yard? _____

4. How many inches are in 5 feet? _____

5. How many centimeters are in 1 meter? _____

6. How many decimeters are in 1 meter? _____

7. Draw a line segment that is about 10 centimeters long.

Now measure it to see how long it really is. _____ centimeters

Measure each line segment to the nearest centimeter.

8.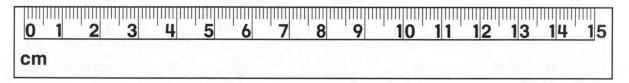

_____

9.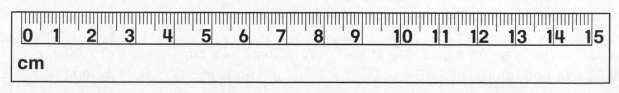

_____

Practice Set 21

Find the perimeter of each figure.

Example

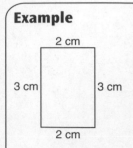

2 cm

3 cm   3 cm

2 cm

Add the lengths of the sides together:

2 cm + 3 cm + 2 cm + 3 cm = **10 cm**

Perimeter = **10 cm**

1.

5 cm

5 cm   5 cm

5 cm

Perimeter: _____

2.

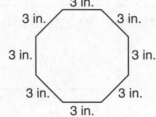

3 in.

2 in.   2 in.

3 in.

Perimeter: _____

3.

7 in.

5 in.

6 in.   4 in.   6 in.

Perimeter: _____

Name each figure. Then find the perimeter.

4.

5 in.   5 in.

5 in.   5 in.

5 in.

_____

Perimeter: _____

5.

3 in.

3 in.   3 in.

3 in.   3 in.

3 in.   3 in.

3 in.

_____

Perimeter: _____

Which figure has the greater perimeter? _____

Use your toolkit clock to help you solve these problems.

6. Jamal leaves to go on a bike ride at 12:30 P.M. and returns at 2:45 P.M. How long was he gone?

_____

7. Christina's party begins at 2:00 P.M. and ends at 5:30 P.M. How long is the party?

_____

Practice Set 22

Find the perimeter of each figure.

1.

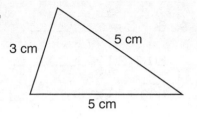

3 cm   5 cm   5 cm

Perimeter: _____

2.

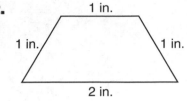

1 in.   1 in.   1 in.   2 in.

Perimeter: _____

3.

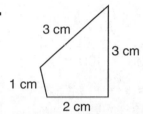

3 cm   3 cm   1 cm   2 cm

Perimeter: _____

4.

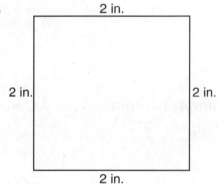

2 in.   2 in.   2 in.   2 in.

Perimeter: _____

5.

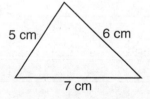

5 cm   6 cm   7 cm

Perimeter: _____

6.

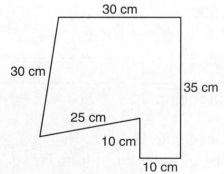

30 cm   30 cm   35 cm   25 cm   10 cm   10 cm

Perimeter: _____

Practice Set 22 *continued*

SRB
50 51
250–253

COMPUTATION PRACTICE **Find each sum or difference.**

7. 7 + 4 = _____

17 + 4 = _____

27 + 4 = _____

37 + 4 = _____

47 + 4 = _____

8. 14 − 6 = _____

24 − 6 = _____

34 − 6 = _____

44 − 6 = _____

54 − 6 = _____

9. 6 + 3 = _____

60 + 30 = _____

600 + 300 = _____

10. 8 − 2 = _____

80 − 20 = _____

800 − 200 = _____

11. 9 + 6 = _____

90 + 60 = _____

900 + 600 = _____

12. 18 − 9 = _____

180 − 90 = _____

1,800 − 900 = _____

13. **Writing/Reasoning** What pattern do you see in your answers to Problem 12?

_____

_____

Count by 100s. Find the missing numbers.

14. 1,200; 1,100; _____; 900; 800; _____; _____; 500; _____; 300

15. 5,630; 5,530; _____; _____; 5,230; 5,130; _____; 4,930; _____; _____

16. 7,659; _____; _____; 7,359; _____; _____; _____; 6,959; _____; _____

Solve each problem.

17. The distance from Dallas to Houston is 245 miles. The distance from Dallas to El Paso is 617 miles. How much farther is it from Dallas to El Paso than from Dallas to Houston?

18. On their vacation, the Baker family drove 376 miles from Phoenix to Los Angeles. Then the Bakers drove 387 miles to San Francisco. How many miles did they drive in all?

_____

_____

Use with or after Lesson 3•5.

Practice Set 23

Find the number of tiles in each shape.

1.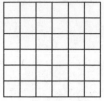

_____

2.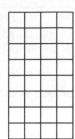

_____

3.

_____

4.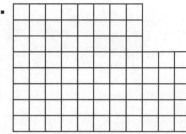

_____

5.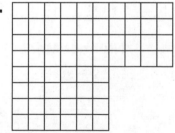

_____

6.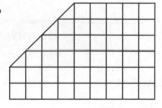

_____

Count by nines. Find the missing numbers.

7. 81, 90, _____, 108, _____, _____, 135, _____, _____, 162

8. 198, _____, _____, 225, 234, 243, _____, _____, _____, 279

9. _____, 612, 621, _____, _____, _____, 657, _____, _____, _____

10. 9, _____, _____, _____, _____, _____, _____, _____, _____, _____

Practice Set 24

SRB
154–156

Write a number model for each rectangle. Then find the area.

Example

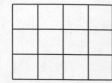

Number model: $3 \times 4 = 12$
Area = **12 square units**

1.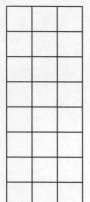

Number Model:

_____

Area:

_____

2.

Number Model:

_____

Area:

_____

3.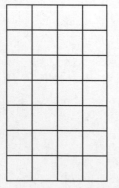

Number Model:

_____

Area:

_____

4.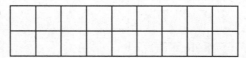

Number Model:

_____

Area:

_____

 Add. Remember to practice and memorize your addition facts.

5. $7 + 8 =$ _____

6. $10 + 9 =$ _____

7. $6 + 5 =$ _____

8. $4 + 3 =$ _____

9. $2 + 1 =$ _____

10. $12 + 0 =$ _____

Use with or after Lesson 3•8.

Practice Set 24 *continued*

Find the perimeter (P) of each figure.

11.

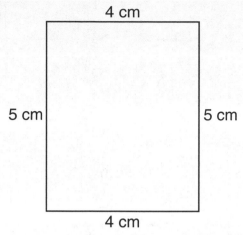

4 cm

5 cm   5 cm

4 cm

P = _____

12.

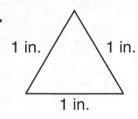

1 in.   1 in.

1 in.

P = _____

13.

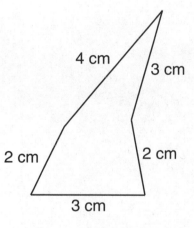

4 cm

3 cm

2 cm   2 cm

3 cm

P = _____

14.

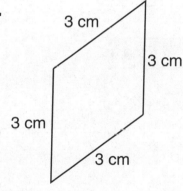

3 cm

3 cm

3 cm

3 cm

P = _____

15. **Writing/Reasoning** If you are building a patio with bricks, will you need to know the area or the perimeter? Explain your answer.

_____

_____

16. **Writing/Reasoning** If you are putting a fence around your yard, will you need to find the area or the perimeter? Explain your answer.

_____

Practice Set 25

The diameter is given. Find the circumference (C) by using the "about 3 times" Circle Rule.

1.

1 in.

C = _____

2.

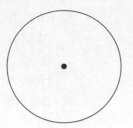

3 cm

C = _____

3.

2 cm

C = _____

Count the bills and coins. Then write the correct amount using dollars-and-cents notation.

4.

_____

5.

_____

6.

_____

7.

_____

Name _____ Date _____ Time _____

Practice Set 26

For Problems 1–4, draw pictures or use counters to help you answer the questions.

1. David has 5 vases. He put 6 flowers in each vase. How many flowers did David put in all of the vases?

_____

2. Sharon bought 4 packs of crackers. Each pack holds 8 crackers. How many crackers did Sharon buy in all?

_____

3. A meeting room has 5 rows of chairs. Each row has 8 chairs. How many chairs are in all of the rows?

_____

4. Each page of a photo album has 4 rows of pictures. Each row has 3 pictures. How many pictures are on each page of the album?

_____

5. **Writing/Reasoning** Mark buys 3 packages of erasers. Each package contains 6 erasers. Can Mark give one eraser to each of the 20 students in his classroom? Explain your answer.

_____

_____

COMPUTATION PRACTICE **Add or subtract.**

6. $112 + 65 =$ _____

7. $197 - 53 =$ _____

8. $116 + 239 =$ _____

9. $456 - 327 =$ _____

10. $272 + 351 =$ _____

11. $923 - 685 =$ _____

12. $49 + 327 + 22 =$ _____

13. $708 - 349 =$ _____

14. $203 + 75 + 81 =$ _____

15.
$$\begin{array}{r} 152 \\ + 398 \\ \hline \end{array}$$

16.
$$\begin{array}{r} 941 \\ - 621 \\ \hline \end{array}$$

17.
$$\begin{array}{r} 384 \\ - 139 \\ \hline \end{array}$$

18.
$$\begin{array}{r} 516 \\ 225 \\ + 394 \\ \hline \end{array}$$

Use with or after Lesson 4·1.

Practice Set 27

Draw or build an array to help you solve each problem.

1. Tyler bought 3 boxes of snacks. Each box had 10 bags of snacks.
   How many bags of snacks did Tyler buy in all?

2. 4 pies are cut into 6 pieces each. How many pieces of pie are there in all?

3. Sharon bought 3 packs of invitations. Each pack had 8 invitations.
   How many invitations did Sharon buy in all?

4. Nancy displays her glass animals in a case that has 5 shelves. Nancy
   puts 4 animals on each shelf. How many animals are in her display
   case?

5. Steve needs 5 inches of ribbon for each puppet that he is making.
   How many inches of ribbon will he need for 8 puppets?

Find the total value of each set of money.

6. 3 Ⓠ 2 Ⓓ 4 Ⓟ _____

7. 1 $1 2 Ⓠ 7 Ⓓ 1 Ⓝ _____

8. 2 $1 4 Ⓠ 3 Ⓓ 4 Ⓝ 3 Ⓟ _____

9. 1 $1 5 Ⓠ 6 Ⓝ 7 Ⓟ _____

10. 2 $1 1 Ⓠ 5 Ⓓ 3 Ⓝ 2 Ⓟ _____

Use with or after Lesson 4·2.

Practice Set 27 *continued*

Find the area (A) of each square or rectangle in square units.

11.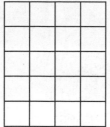

A = _____

12.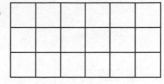

A = _____

13.

A = _____

14.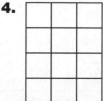

A = _____

15.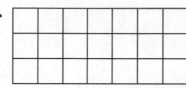

A = _____

16.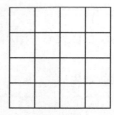

A = _____

17. **Writing/Reasoning** Find the area of a grid that is 8 units tall and 6 units wide. Draw a picture and explain your work.

_____

_____

Practice Set 28

For Problems 1–6, use counters to help you solve the division problems.

24 grapes shared equally …

1. *by 3 people*

_____ grapes
per person

_____ grapes
left over

2. *by 4 people*

_____ grapes
per person

_____ grapes
left over

3. *by 6 people*

_____ grapes
per person

_____ grapes
left over

48 cherries shared equally …

4. *by 6 people*

_____ cherries
per person

_____ cherries
left over

5. *by 8 people*

_____ cherries
per person

_____ cherries
left over

6. *by 12 people*

_____ cherries
per person

_____ cherries
left over

Subtract. Remember to practice and memorize your subtraction facts.

7. $89 - 67 =$ _____

8. $58 - 23 =$ _____

9. $90 - 36 =$ _____

10. $32 - 18 =$ _____

11. $77 - 56 =$ _____

12. $46 - 21 =$ _____

13. $\begin{array}{r} 73 \\ -\ 56 \\ \hline \end{array}$

14. $\begin{array}{r} 53 \\ -\ 29 \\ \hline \end{array}$

15. $\begin{array}{r} 62 \\ -\ 26 \\ \hline \end{array}$

16. $\begin{array}{r} 63 \\ -\ 48 \\ \hline \end{array}$

17. $\begin{array}{r} 84 \\ -\ 35 \\ \hline \end{array}$

18. $\begin{array}{r} 75 \\ -\ 28 \\ \hline \end{array}$

19. $\begin{array}{r} 48 \\ -\ 19 \\ \hline \end{array}$

20. $\begin{array}{r} 92 \\ -\ 53 \\ \hline \end{array}$

Use with or after Lesson 4·3.

Practice Set 29

For each problem, write a number model. Then find
the missing numbers.

> **Example** 31 apples are divided equally among 6 baskets.
> How many apples are in each basket?
>
> **Number model: 31 ÷ 6 → 5 R1**
>
> **Each basket has 5 apples.**
> **1 apple is left over.**

1. 24 bones are shared
equally among 6 dogs.
How many bones
does each dog get?

_____ ÷ _____ → _____ R _____

Each dog gets _____ bones.

_____ bones are left over.

2. Tim has 27 jars of jam.
He puts 4 jars in each
box. How many boxes
does he fill?

_____ ÷ _____ → _____ R _____

Tim fills _____ boxes.

_____ jars are left over.

Find the missing numbers.

3.

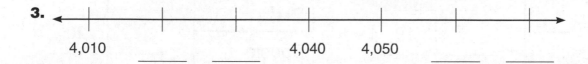

4,010 _____ _____ 4,040 4,050 _____ _____

4.

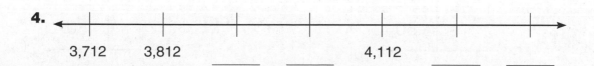

3,712 3,812 _____ _____ 4,112 _____ _____

5.

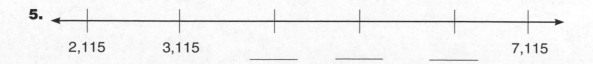

2,115 3,115 _____ _____ _____ 7,115

Practice Set 29 *continued*

Find the missing numbers.

6.

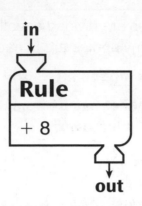

in	out
12	8
14	10
	13
	15
	21

in
Rule
− 4
out

7.

in	out
	15
	12
	20
	32
	40

in
Rule
+ 8
out

8.

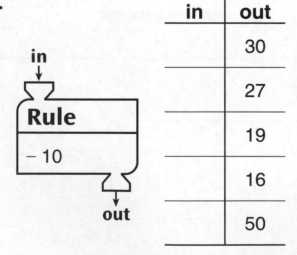

in	out
	30
	27
	19
	16
	50

in
Rule
− 10
out

9.

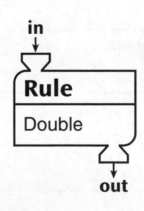

in	out
	8
	14
	20
	24
	50

in
Rule
Double
out

Use with or after Lesson 4•4.

Practice Set 30

Solve each multiplication problem. Then write a turn-around shortcut for each problem.

Example $6 \times 2 = 12$ $2 \times 6 = 12$

1. $3 \times 2 = $ _____

2. $4 \times 3 = $ _____

3. $3 \times 5 = $ _____

_____

_____

_____

4. $6 \times 3 = $ _____

5. $7 \times 4 = $ _____

6. $2 \times 4 = $ _____

_____

_____

_____

 Multiply. Remember to practice and memorize your multiplication facts.

7. $5 \times 0 = $ _____

8. $7 \times 1 = $ _____

9. $3 \times 1 = $ _____

10. $16 \times 1 = $ _____

11. $0 \times 4 = $ _____

12. $12 \times 0 = $ _____

13. $10 \times 0 = $ _____

14. $6 \times 1 = $ _____

15. $2 \times 0 = $ _____

Write each amount with dollars and cents.

Example $1 $1   Q D N N   **$2.45**

16. $1 Q Q D D P P P _____

17. $1 $1 $1 Q D D N P P _____

18. $10 $1 D D D D N P _____

19. $1 Q Q Q Q Q D P P P P _____

Practice Set 31

Find the missing number for each Fact Triangle.
Then write the family of facts for that triangle.

Example

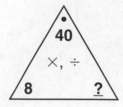

Missing number: 5
Fact family:
8 × 5 = 40
5 × 8 = 40
40 ÷ 8 = 5
40 ÷ 5 = 8

1.

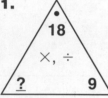

Missing number: _____
Fact family:
_____
_____
_____
_____

2.

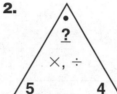

Missing number: _____
Fact family:
_____
_____
_____
_____

3.

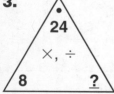

Missing number: _____
Fact family:
_____
_____
_____
_____

4.

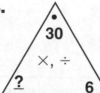

Missing number: _____
Fact family:
_____
_____
_____
_____

5.

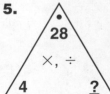

Missing number: _____
Fact family:
_____
_____
_____
_____

6.

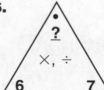

Missing number: _____
Fact family:
_____
_____
_____
_____

Use with or after Lesson 4•6.

Practice Set 31 continued

SRB
35
73 74

Match each amount of money with an equal amount from the list at the right. Then write the letter that identifies that amount.

7. fourteen dollars and two cents _____

    A. twelve dollars and forty cents

8. $20.14 _____

    B. $12.04

9. $41.20 _____

    C. forty-one dollars and twenty cents

10. $12.40 _____

    D. $1.42

11. $10.42 _____

    E. twenty dollars and fourteen cents

12. twelve dollars and four cents _____

    F. ten dollars and forty-two cents

13. one dollar and forty-two cents _____

    G. $14.02

Find the missing numbers. You can use counters or draw pictures.

14. 15 pieces of candy
    4 children share equally

    _____ pieces per child

    _____ pieces remaining

15. 12 tennis balls
    3 balls per can

    _____ filled cans

    _____ balls remaining

16. 14 carrots
    6 rabbits share equally

    _____ carrots per rabbit

    _____ carrots remaining

17. 27 books
    8 books per box

    _____ filled boxes

    _____ books remaining

Practice Set 32

Find the missing number for each Fact Triangle.
Then write the family of facts for that triangle.

1.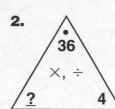
48
×, ÷
6   ?

Missing number: _____
Fact family:

_____

_____

_____

2.
36
×, ÷
?   4

Missing number: _____
Fact family:

_____

_____

_____

3. Taylor has 3 sheets of stickers. Each sheet has 10 stickers. How many stickers does he have in all?

Answer the question: _____

Number model: _____

Fact family:

_____

_____

_____

_____

4. Manuel practices his trumpet for 15 minutes each day. How many minutes does he practice in 4 days?

Answer the question: _____

Number model: _____

Fact family:

_____

_____

_____

_____

COMPUTATION PRACTICE Add or subtract. Remember to practice and memorize your basic facts.

5. $398 + 64 =$ _____

6. $59 + 211 =$ _____

7. $600 - 75 =$ _____

8. $282 - 139 =$ _____

Use with or after Lesson 4•7.

Practice Set 33

Multiply the number in the *center* of the circle by each number *on* the circle. Then write the product *outside* the circle.

Example

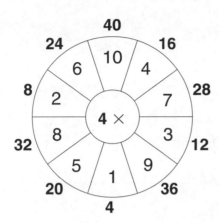

1.

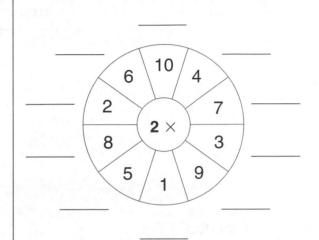

2.

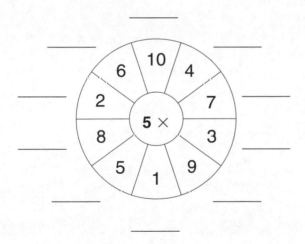

3.

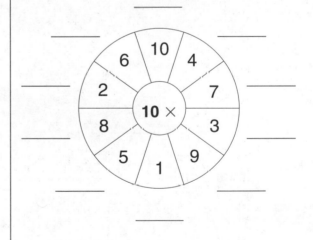

Write the value of the underlined digit in each number.

Example   5,416   **5,000** or 5 thousands

4. 7<u>9</u>6 _____

5. 7,51<u>4</u> _____

6. <u>6</u>48 _____

7. 8,<u>9</u>54 _____

8. <u>4</u>2,597 _____

9. 9,0<u>4</u>6 _____

10. 58,0<u>2</u>9 _____

11. <u>6</u>,309 _____

12. 3,1<u>8</u>9 _____

Practice Set 33 continued

Write a multiplication fact to find the total number of dots in each array.

Example	: : : : : : :	$3 \times 7 = 21$
	: : : : : : :	21 total dots
	: : : : : : :	

13. : : : :
: : : :
: : : :
: : : :

_____ × _____ = _____

_____ total dots

14. : : : :
: : : :
: : : :
: : : :
: : : :

_____ × _____ = _____

_____ total dots

15. : :
: :
: :
: :
: :

_____ × _____ = _____

_____ total dots

16. : : :
: : :
: : :
: : :
: : :

_____ × _____ = _____

_____ total dots

17. : : : : : :
: : : : : :

_____ × _____ = _____

_____ total dots

18. : : : : :
: : : : :
: : : : :
: : : : :

_____ × _____ = _____

_____ total dots

Use with or after Lesson 4·8.

Practice Set 34

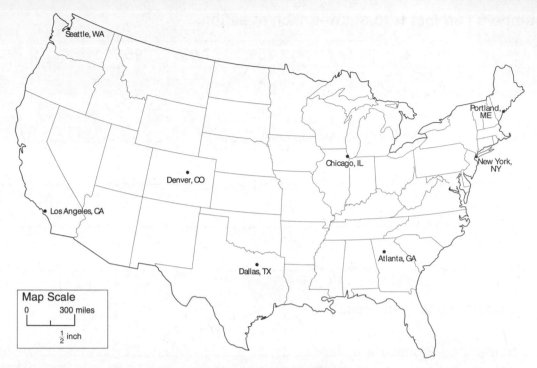

Use the map and the map scale to find the distances.

1. The distance between New York and Atlanta is about _____ inches on the map. That is about _____ miles.

2. The distance between Dallas and New York is about _____ inches on the map. That is about _____ miles.

3. The distance between Seattle and Los Angeles is about _____ inches on the map. That is about _____ miles.

4. The distance between Denver and Chicago is about _____ inches on the map. That is about _____ miles.

5. The distance between Los Angeles and Portland is about _____ inches on the map. That is about _____ miles.

Write <, >, or = to compare the numbers.

6. 125 _____ 122

7. 65 _____ 83

8. 1,211 _____ 3,289

9. 56 _____ 43

Practice Set 35

Use the numbers in the box to answer each question.

| 79,512 | 29,517 | 12,759 | 27,951 | 95,721 |

1. Which numbers have 5 tens? _____

2. Which number has exactly 5 thousands? _____

3. Which numbers have exactly 1 ten? _____

4. Which numbers have 9 thousands? _____

5. Which number has 9 ten-thousands? _____

6. Which numbers have 7 hundreds? _____

7. Which number has 2 ones? _____

8. Which numbers have 5 hundreds? _____

9. Which numbers have 2 ten-thousands? _____

10. Which numbers have 1 one? _____

11. Which number has 7 thousands? _____

 Multiply. Remember to practice and memorize your multiplication facts.

12. $1 \times 7 =$ _____

13. $5 \times 2 =$ _____

14. $8 \times 6 =$ _____

15. $8 \times 2 =$ _____

16. $3 \times 10 =$ _____

17. $5 \times 7 =$ _____

18. $6 \times 5 =$ _____

19. $9 \times 5 =$ _____

20. $4 \times 9 =$ _____

21. $10 \times 5 =$ _____

22. $8 \times 0 =$ _____

23. $6 \times 1 =$ _____

24. $3 \times 2 =$ _____

25. $6 \times 6 =$ _____

26. $5 \times 7 =$ _____

27. $4 \times 4 =$ _____

28. $8 \times 4 =$ _____

29. $6 \times 3 =$ _____

Use with or after Lesson 5·1.

Practice Set 35 continued

Find the missing rule and the numbers for the empty frames.

Example

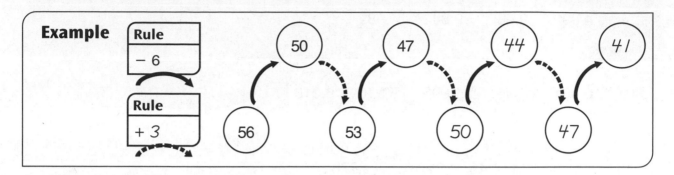

30.

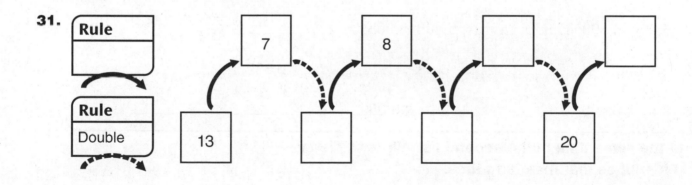

31.

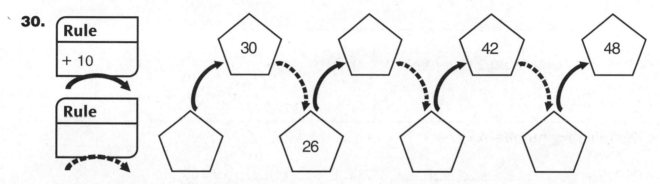

32.

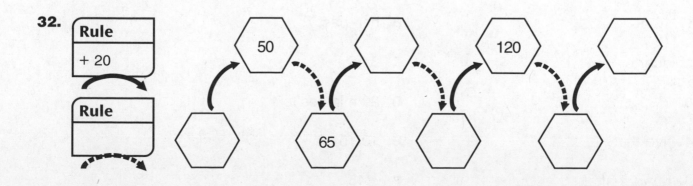

Practice Set 36

Write each group of numbers in order from least to greatest.

1. 4,289   4,892   2,489   9,842   4,982

_____

2. 5,901   6,001   5,991   1,995   5,910   6,010

_____

3. 10,453   1,543   11,246   21,101   9,878

_____

4. 71,034   17,340   80,249   99,999   81,001

_____

Write the missing numbers.

5. 2,300; 2,400; _____; _____; _____; 2,800; _____; _____; 3,100; 3,200

6. 7,260; 7,360; _____; _____; _____; 7,760; _____; _____; 8,060; _____

7. 5,408; 5,508; _____; _____; 5,808; 5,908; _____; _____; _____

8. 6,132; _____; _____; _____; 6,532; _____; _____; 6,832; _____; _____

9. 8,555; 8,655; _____; _____; _____; 9,055; _____; _____; _____

**Find the same time in the second list. Write the letter
that identifies that matching time.**

10. 5 minutes after 3 _____   **A.** 5:30

11. 10:40 _____   **B.** half-past 7

12. quarter-to 5 _____   **C.** 3:05

13. 7:30 _____   **D.** 20 minutes to 11

14. five-thirty _____   **E.** 10:15

15. quarter-after 10 _____   **F.** 4:45

64

Practice Set 37

Write the following numbers by using digits:

1. one million, two hundred twenty-eight thousand _____

2. six million, three thousand, four hundred _____

3. seven hundred thirty-one thousand, five hundred forty-nine _____

4. eighty-three thousand, nine hundred two _____

5. four million, six hundred five _____

6. three million, twenty thousand, five hundred _____

Write the following numbers by using words:

7. 5,007,023 _____

8. 603,401 _____

 Add or subtract.

9. 721 − 350 = _____   **10.** 213 + 643 = _____   **11.** 672 − 514 = _____

12. 815 + 192 = _____   **13.** 728 − 456 = _____   **14.** 359 + 287 = _____

15. 821
      − 371

16. 416
      − 203

17.   89
      + 376

18. 327
      − 119

19. 223
      + 478

20. 632
      − 218

21. 551
       321
      + 114

22. 264
       118
      + 319

Practice Set 38

Write < or >.

> | Example | 59,423 _____ 59,389 |
> | | Both numbers have 5 ten-thousands and 9 thousands. |
> | | 4 hundreds is greater than 3 hundreds. |
> | | Therefore: 59,423 > 59,389 |

> < means *is less than*
> > means *is greater than*

1. 127,675 _____ 137,675

2. 24,714 _____ 24,710

3. 159,338 _____ 160,273

4. 673,218 _____ 673,239

5. 285,641 _____ 385,641

6. 490,315 _____ 510,214

7. 331,846 _____ 330,259

8. 37,014 _____ 37,104

9. 999,972 _____ 999,992

10. 53,892 _____ 54,617

For each number below, write the number that is 10 more, 100 more, and 1,000 more.

> | Example | 33,492 | 10 more: **33,502** |
> | | | 100 more: **33,592** |
> | | | 1,000 more: **34,492** |

11. 52,416

12. 68,927

13. 72,499

_____

_____

_____

14. 65,798

15. 95,281

16. 39,482

_____

_____

_____

Practice Set 39

Write < or > to compare the numbers.

1. 3,845,230 _____ 3,856,400

2. 9,457,432 _____ 9,145,491

Write these numbers in order from least to greatest.

3. 3,654,349   2,425,569   3,549,902   2,459,340

_____

4. 1,005,402   948,340   1,240,341   1,050,397

_____

5. ✏️ **Writing/Reasoning** Write a number story using a 7-digit number.

_____

_____

_____

_____

_____

Write the value of the underlined digit in each number.

6. 3<u>7</u>,450 _____   **7.** 125,<u>8</u>92 _____

Write the number that has ...

8. 6 hundreds
9 ones
3 ten-thousands
4 tens
8 thousands

_____

9. 5 thousands
9 tens
9 hundreds
2 ten-thousands
8 ones
3 hundred-thousands

_____

Practice Set 40

What number is shown by the base-10 blocks?

1. 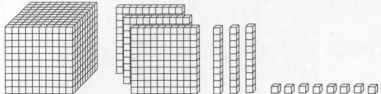 _____

2. _____

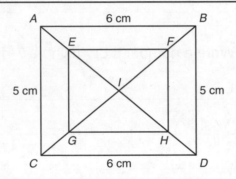

3. What shape is *IFH above*? _____

4. How many different triangles are in the figure? _____

5. What is the perimeter of *ABCD*? _____

6. ✎ **Writing/Reasoning** Describe how to find the area of rectangle *ABCD*.

_____

_____

Find each missing number.

7. 2 meters = _____ centimeters

8. _____ decimeters = 40 centimeters

9. 300 centimeters = _____ meters

10. 400 decimeters = _____ meters

Practice Set 41

Write a decimal for the shaded part of each grid. Each grid is ONE.

Example

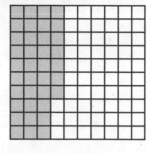

0.37
_____

1.

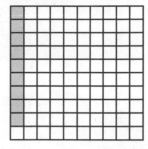

_____

2.

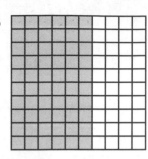

_____

3.

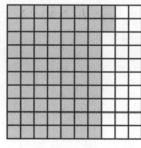

_____

4.

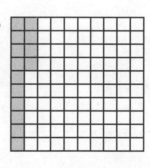

_____

5.

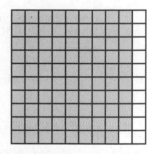

_____

6.

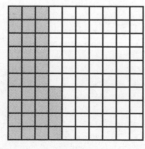

_____

7.

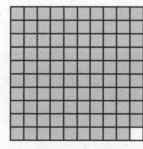

_____

8.

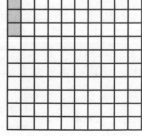

_____

Practice Set 41 continued

Write each number by using digits.

> **Example** twenty-six thousand, four hundred twelve
> **26,412**

9. four thousand, six hundred ten _____

10. seventy-two thousand, eight hundred five _____

11. five thousand, six _____

12. sixty-seven thousand, three hundred eighteen _____

13. one hundred fourteen thousand, five hundred thirty-six _____

14. seven hundred eighty thousand, two hundred _____

15. six hundred seventy-nine thousand, eight _____

16. four hundred three thousand, ninety-two _____

Use your calculator. Write the answer in dollars and cents.

> **Example** $2.71 + 92¢ = **$3.63**

17. 67¢ + $1.21 = _____

18. $0.59 + 83¢ + 27¢ = _____

19. $3.41 + $1.08 = _____

20. 114¢ + 32¢ = _____

21. $0.65 + $0.84 = _____

22. $5.04 + 42¢ + 33¢ = _____

23. $0.84 + $2.86 = _____

24. 256¢ + $2.56 = _____

25. 49¢ + $1.92 + 67¢ = _____

26. $4.41 + $3.15 + $6.85 = _____

Practice Set 42

Use the decimals in the box to answer each question below.

1. Which number has exactly 3 tenths? _____

2. Which number has exactly 5 hundredths? _____

3. Which number has exactly 7 tenths? _____

4. Which number has 9 hundredths? _____

5. Which number has 0 tenths? _____

6. Which number has 6 tenths and 7 hundredths? _____

7. Which number has 5 tenths and 6 hundredths? _____

8. Which number has 3 tenths and 4 hundredths? _____

0.03
0.34
0.45
0.56
0.67
0.78
0.89

9. **Writing/Reasoning** Tell why the decimal numbers 2.03 and 2.30 are not equal. Use words, numbers, or pictures to explain your answer.

Write the addition and subtraction fact family for each group of numbers.

Example	5, 6, 11
	5 + 6 = 11
	6 + 5 = 11
	11 − 6 = 5
	11 − 5 = 6

10. 2, 8, 10

11. 7, 8, 15

12. 9, 3, 12

_____   _____   _____

_____   _____   _____

_____   _____   _____

_____   _____   _____

Practice Set **42** *continued*

For each number, write the number that is 10 less, 100 less, and 1,000 less.

Example	52,928
	10 less: **52,918**
	100 less: **52,828**
	1,000 less: **51,928**

13. 27,386

_____

_____

14. 50,221

_____

_____

15. 38,482

_____

_____

16. 93,525

_____

_____

17. 27,058

_____

_____

18. 60,867

_____

_____

19. 62,505

_____

_____

_____

20. 28,331

_____

_____

_____

21. 83,471

_____

_____

_____

Solve each problem.

22. David had 527 baseball cards in his collection. He got 85 more baseball cards for his birthday. How many baseball cards does he have now?

_____

23. Jenny has 412 hockey cards and 843 basketball cards. How many more basketball cards than hockey cards does she have?

_____

Practice Set 43

Write <, >, or =.

■ ←cube (1 cm)

▬▬▬ ←long (10 cm)

▬▬▬▬▬▬▬▬▬▬▬▬▬▬▬▬▬▬▬▬ ←meterstick

1. 0.01 meter _____ 0.08 meter

2. 0.30 meter _____ 0.03 meter

3. 0.40 meter _____ 0.20 meter

4. 0.50 meter _____ 0.54 meter

5. 0.1 meter _____ 2 longs

6. 0.01 meter _____ 5 cubes

7. 0.07 meter _____ 7 cubes

8. 0.6 meter _____ 6 cubes

Write the multiplication and division fact family for each group of numbers.

Example   16, 8, 2

$$8 \times 2 = 16$$
$$2 \times 8 = 16$$
$$16 \div 2 = 8$$
$$16 \div 8 = 2$$

9. 4, 5, 20

10. 2, 4, 8

11. 3, 4, 12

_____

_____

_____

_____

12. 5, 2, 10

13. 9, 2, 18

14. 5, 6, 30

_____

_____

_____

Practice Set 44

SRB
137
190–192

Use the ruler to answer the following questions.

```
0  1  2  3  4  5  6  7  8  9  10  11  12  13  14  15
cm
```

1. How many millimeters are in a centimeter? _____

2. How many centimeters are in a meter? _____

3. How many millimeters are in 2 centimeters? _____

4. How many millimeters are in a meter? _____

5. How many millimeters are in 24.3 centimeters? _____

Make a ballpark estimate to answer each question. Then write *yes* or *no*.

6. You have $9.00. Do you have enough to buy a book for $3.60, a magazine for $2.29, and a poster for $2.79?

   Ballpark estimate:

   _____

   Answer: _____

7. You have $12.00. Do you have enough to buy a beach towel for $7.30 and sunglasses for $4.45?

   Ballpark estimate:

   _____

   Answer: _____

8. You have $20.00. Do you have enough to buy a shirt for $11.40 and shorts for $7.39?

   Ballpark estimate:

   _____

   Answer: _____

9. You have $15.00. Do you have enough to buy a fishing pole for $9.50, fishing line for $3.89, and bait for $1.79?

   Ballpark estimate:

   _____

   Answer: _____

Use with or after Lesson 5•10.

Practice Set 44 continued

Find each missing number.

1 m = 10 dm	1 dm = 0.1 m
1 m = 100 cm	1 cm = 0.01 m
1 dm = 10 cm	1 cm = 0.1 dm

10. _____ m = 400 cm

11. 0.8 dm = _____ cm

12. 60 cm = _____ dm

13. 0.5 m = _____ dm

14. _____ dm = 9 m

15. _____ cm = 0.02 m

16. 7 m = _____ cm

17. 0.4 m = _____ dm

18. 8 cm = _____ m

19. _____ m = 9 cm

20. _____ cm = 3 m

21. _____ cm = 0.9 dm

Find the missing numbers.

22.

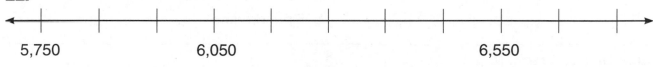

5,750   6,050   6,550

23.

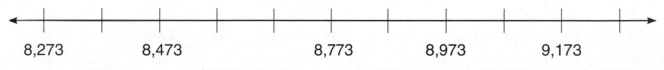

8,273   8,473   8,773   8,973   9,173

FACTS PRACTICE **Multiply. Remember to practice and memorize your multiplication facts.**

24. 7 × 8 = _____   **25.** 9 × 6 = _____   **26.** 6 × 5 = _____   **27.** 7 × 2 = _____

28. 3 × 9 = _____   **29.** 4 × 8 = _____   **30.** 9 × 9 = _____   **31.** 3 × 3 = _____

32. 5 × 9 = _____   **33.** 6 × 8 = _____   **34.** 7 × 4 = _____   **35.** 5 × 3 = _____

36. 4 × 6 = _____   **37.** 8 × 8 = _____   **38.** 2 × 6 = _____   **39.** 9 × 8 = _____

Practice Set 45

Match each number in the first list with the same number in the
second list. Write the letter that identifies that matching number.

1. 0.5 _____   **A.** 0.004

2. two hundred one thousandths _____   **B.** 2 tenths

3. 0.521 _____   **C.** 0.040

4. 0.05 _____   **D.** 5 tenths

5. 21 thousandths _____   **E.** 21 hundredths

6. 4 thousandths _____   **F.** 0.201

7. 0.4 _____   **G.** 521 thousandths

8. 0.2 _____   **H.** 0.021

9. 40 thousandths _____   **I.** 4 tenths

10. 0.21 _____   **J.** 5 hundredths

Write < or > for each.

11. 465,243 _____ 564,243   **12.** 107,453 _____ 107,452

13. 999,999 _____ 1,000,000   **14.** 382,591 _____ 382,491

15. 848,484 _____ 484,848   **16.** 12,495 _____ 112,495

17. 359,416 _____ 369,416   **18.** 992,450 _____ 992,460

19. 600,770 _____ 600,707   **20.** 739,418 _____ 843,291

21. 57,989 _____ 56,989   **22.** 199,998 _____ 200,000

23. 632,521 _____ 6,328   **24.** 12,345 _____ 21,345

25. 489,201 _____ 489,221   **26.** 5,891 _____ 15,891

Practice Set 46

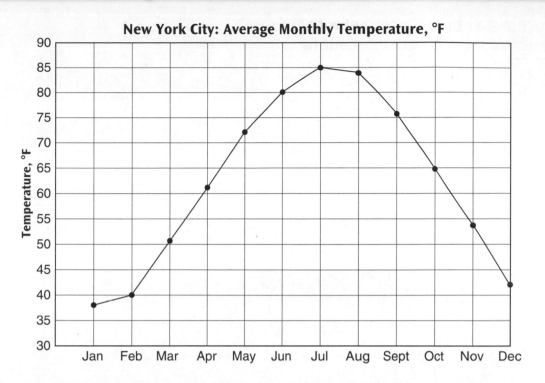

New York City: Average Monthly Temperature, °F

Use the data on the line graph to answer the questions below.

1. What is the average monthly temperature in February? _____

2. What is the average monthly temperature in April? _____

3. Which month is the least warm on average? _____

4. Which two months have the highest average temperatures? _____

5. **Writing/Reasoning** What is the range of temperatures? Explain how you found your answer.

_____

_____

6. **Writing/Reasoning** Describe the trend in the data that is shown by the graph.

_____

_____

Practice Set 47

SRB
102
52 53

Count the number of line segments used to make each figure.

1.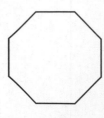

   _____

2.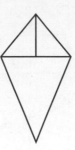

   _____

3.

   _____

4.

   _____

5.

   _____

6.

   _____

For each problem below, write a number model.
Then write the number to complete each sentence.

7. A group of 4 children share 12 pens.
   How many pens does each child get?

   _____ ÷ _____ → _____ R _____

   Each child gets _____ pens.

   _____ pens are left over.

8. Three cats share 8 toy mice. How many
   mice does each cat get?

   _____ ÷ _____ → _____ R _____

   Each cat gets _____ mice.

   _____ mice are left over.

9. Brian has 11 sweaters and puts 3 in
   each drawer. How many drawers does
   Brian fill?

   _____ ÷ _____ → _____ R _____

   Brian fills _____ drawers.

   _____ sweaters are left over.

10. Jan buys 10 pencils and shares them
    equally with 3 friends. How many pencils
    does each friend get?

    _____ ÷ _____ → _____ R _____

    Each friend gets _____ pencils.

    _____ pencils are left over.

Use with or after Lesson 6•1.

Practice Set 48

For Problems 1–4, match each description with the correct example. Write the letter that identifies that example.

1. parallel lines _____

A.

2. intersecting lines _____

B.

3. intersecting line segments _____

C.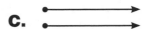

4. parallel rays _____

D.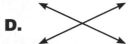

5. Draw a pair of parallel line segments.

6. Draw a pair of intersecting rays.

Write the multiplication and division fact family for each group of numbers.

7. 25, 5, 5

_____

_____

8. 2, 4, 2

_____

_____

9. 8, 64, 8

_____

_____

10. 9, 9, 81

_____

_____

11. 36, 6, 6

_____

_____

12. 7, 7, 49

_____

_____

13. 8, 72, 9

_____

_____

_____

14. 6, 30, 5

_____

_____

_____

15. 27, 9, 3

_____

_____

_____

Practice Set 48 continued

Measure each object to the nearest half-inch or half-centimeter.

16.

_____ cm

17.

_____ in.

18.

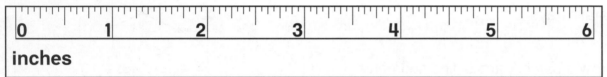

_____ cm

19.

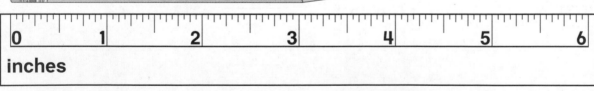

_____ in.

 Add or subtract. Remember to practice and memorize your basic facts.

20.	210	**21.**	459	**22.**	234	**23.**	982
	+ 354		+ 863		− 158		− 491

Use with or after Lesson 6·2.

Practice Set 49

The shaded part of each clock shows passing time.
Assume that each clock turns clockwise.

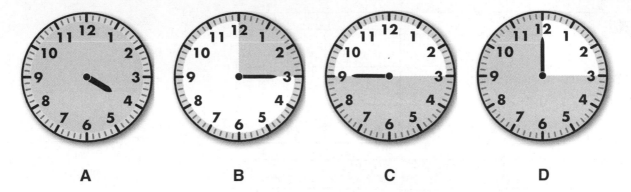

| A | B | C | D |

1. Which clock shows that one hour has passed? _____

2. Which clock shows that forty-five minutes have passed? _____

3. Which clock shows that half an hour has passed? _____

4. Which clock shows that 15 minutes have passed? _____

Which clock shows ...

5. a full turn? _____   6. a half-turn? _____

7. a quarter-turn? _____   8. a $\frac{3}{4}$ turn? _____

Draw an array to find each product.

9. $4 \times 5 =$ _____   10. $3 \times 8 =$ _____   11. $9 \times 4 =$ _____

12. $2 \times 7 =$ _____   13. $5 \times 5 =$ _____   14. $6 \times 1 =$ _____

Practice Set 49 *continued*

Find the area of each rectangle in square centimeters.
Then find the perimeter of each rectangle in centimeters.

Example

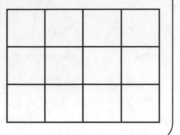

Area: **12 square centimeters**
Perimeter: **14 centimeters**

15.

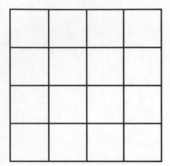

Area: _____

Perimeter: _____

16.

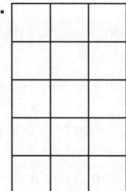

Area: _____

Perimeter: _____

17.

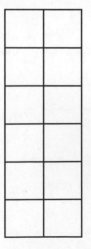

Area: _____

Perimeter: _____

18.

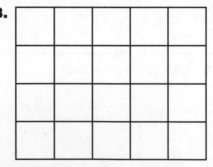

Area: _____

Perimeter: _____

Use with or after Lesson 6•3.

Practice Set 50

SRB
103 104
108 109

Use the quadrangles to answer Problems 1–6.

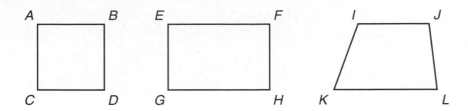

1. All of the angles in *ABCD* and *EFGH* are _____ and the

   sides of *ABCD* are _____ in length.

2. Quadrangle *ABCD* is a _____.

3. Quadrangle *EFGH* is a _____.

4. Quadrangle *IJLK* is a _____.

5. How is quadrangle *EFGH* different from quadrangle *IJKL*?

   _____

6. [✎] **Writing/Reasoning** Is a square always a rectangle?
   Explain your answer.

   _____

   _____

   _____

Write the time shown on each clock.

7.

8.

9.

_____ _____ _____

Practice Set 51

1. Circle the shapes that have right angles.

A.

B.

C.

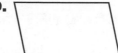

D.

E.

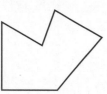

F.

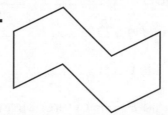

Write the number.

Example	4 tens
	9 hundreds
	0 thousands
	3 ten-thousands
	7 ones **30,947**

2. 8 hundreds
9 ones
5 ten-thousands
3 tens
9 thousands

_____

3. 4 thousands
6 tens
1 hundred
2 ones
7 ten-thousands

_____

4. 7 hundreds
0 ones
3 ten-thousands
4 thousands
9 tens

_____

5. 8 thousands
5 tens
2 hundreds
6 ten-thousands
8 ones

_____

Use with or after Lesson 6•6.

Practice Set 51 continued

Match each description with the correct polygon below.
Write the letter of that polygon.

6. a rectangle with a perimeter of 22 in. _____

7. a triangle with a perimeter of 18 in. _____

8. a parallelogram with a perimeter of 18 in. _____

9. a square with a perimeter of 16 in. _____

10. a kite with a perimeter of 18 in. _____

11. a triangle with a perimeter of 17 in. _____

12. a rhombus with a perimeter of 28 in. _____

13. a rectangle with a perimeter of 20 in. _____

A. 4 in.

B. 5 in. / 5 in. / 7 in.

C. 4 in. / 7 in.

D. 6 in. / 4 in. / 8 in.

E. 4 in. / 5 in.

F. 7 in.

G. 4 in. / 6 in.

H. 7 in. / 2 in.

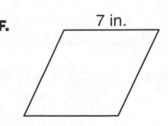

Practice Set 52

Draw each angle as described below. Record the direction
of each turn with a curved arrow. Mark any right angle
you make with a □.

Example An angle that shows a $\frac{3}{4}$ turn

1. An angle that shows a quarter-turn

2. An angle that shows a half-turn

3. An angle that is smaller than a half-turn

4. An angle that is larger than a half-turn

Draw an array to find each product.

Example 3×7 • • • • • • • **21**
  • • • • • • •
  • • • • • • •

5. $2 \times 5 =$ _____

6. $6 \times 4 =$ _____

7. $1 \times 8 =$ _____

8. $9 \times 3 =$ _____

9. $5 \times 6 =$ _____

10. $8 \times 3 =$ _____

11. $5 \times 1 =$ _____

12. $7 \times 6 =$ _____

Use with or after Lesson 6•7.

Practice Set 53

Each picture below shows one-half of a letter. The dashed line is the line of symmetry. Write the complete letter.

1.

2.

3.

_____   _____   _____

4.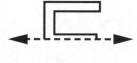

5.

6. Draw another letter that has a line of symmetry.

_____   _____   _____

Draw the number of lines of symmetry shown in parentheses.

7. (3)

8. (1)

9. (1)

Practice Set 53 continued

Cross out the names that DO NOT belong in each name-collection box.
Then write the number that belongs on the label of each box.

Example

21

5 + 6 + 5 + 6

16 + 5   30 − 12

10 + 5 + 6

18 + 2

twenty-one

8 × 3   10 + 11

10.

4 × 8   15

3 × 10   7   + 15

20 + 13   × 4

8 + 9 + 13

5 more than 25

5 × 6   3 × 9

11.

8 × 8   32

9 × 7   + 32

20 + 20 + 20

72 − 8   sixty-one

9 more than 56

100 − 36   46 + 18

12.

20 − 2   9

• • • • • • • • •   + 9
• • • • • • • • •

4 + 5 + 9   2 × 9

2 more than 15

𝖧𝖧𝖳 𝖧𝖧𝖳 𝖧𝖧𝖳 /   8

4 less than 23   + 9

13.

2 × 20   5

9 × 5   × 8

6 less than 45

10 + 10 + 10 + 10

40 × 0   forty-one

14 + 26

5 more than 35

14.

20 less than 60   25

14 + 36   10   + 25

6 × 10   × 5

20 + 20 + 10

0 × 50

30 + 25   100 − 5

15. Look at the shaded squares
on the grid.

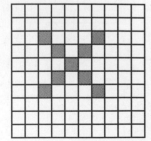

What decimal number do
the shaded squares show?

_____

16. Shade (color in) 34 squares
on this grid.

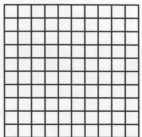

What decimal number do
the shaded squares show?

_____

Use with or after Lesson 6·10.

Practice Set 54

SRB
50 51
112–120

Write the name of each solid. Tell how many faces, edges, and vertices each solid has.

Example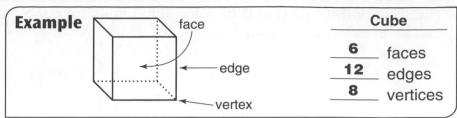

face

edge

vertex

Cube	
6	faces
12	edges
8	vertices

1.
_____
_____ faces
_____ edges
_____ vertices

2.
_____
_____ faces
_____ edges
_____ vertices

3.
_____
_____ faces
_____ edges
_____ vertices

4.
_____
_____ faces
_____ edges
_____ vertices

5.
_____
_____ faces
_____ edges
_____ vertices

6. **Writing/Reasoning** Describe how a cylinder is like a prism. Then tell how it is different from a prism.

_____

_____

_____

COMPUTATION PRACTICE Add. Remember to practice and memorize your addition facts.

7. 42 + 15 = _____

8. 31 + 29 = _____

9. 52 + 64 = _____

10. 37 + 61 = _____

11. 83 + 18 = _____

12. 34 + 68 = _____

13. 72 + 87 = _____

14. 56 + 25 = _____

15. 19 + 31 = _____

Practice Set 54 continued

Write the multiplication and division fact family for each Fact Triangle.

Example

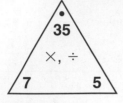

7 × 5 = 35
5 × 7 = 35
35 ÷ 7 = 5
35 ÷ 5 = 7

16.
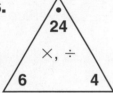
_____

_____

_____

17.
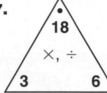
_____

_____

_____

18.

_____

_____

_____

19.
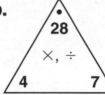
_____

_____

_____

Write multiplication and division facts for each group of numbers.

Example   5, 5, 25   5 × 5 = 25
                               25 ÷ 5 = 5

20. 3, 3, 9 _____

21. 16, 4, 4 _____

22. 6, 36, 6 _____

23. 49, 7, 7 _____

24. 9, 81, 9 _____

25. 8, 64, 8 _____

Use with or after Lesson 6•11.

Practice Set 55

Sandy is making a design by pressing the bases of pyramids and prisms onto an ink pad. What shape can she make from each block?

Example

The base of a triangular prism makes a ___*triangle*___.

1.

_____

2.

_____

3.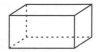

_____

4.

_____

5.

_____

6.

_____

Find the missing numbers. Count back by 1,000s.

7. 10,000; 9,000; _____; _____; _____; 5,000; _____; _____

8. 12,800; 11,800; _____; 9,800; _____; _____; _____; 5,800

9. 13,420; 12,420; _____; _____; _____; 8,420; _____; _____

10. 16,059; 15,059; _____; _____; _____; 11,059; _____; _____

11. 14,955; 13,955; _____; _____; _____; _____; 8,955; _____

Practice Set 55 continued

Write the number.

> **Example** 2 in the tenths place
> 9 in the ones place
> 6 in the tens place
> 8 in the hundredths place   **69.28**

12. 0 in the ones place
4 in the tenths place
8 in the tens place
9 in the hundredths place

_____

13. 8 in the thousandths place
2 in the tenths place
6 in the ones place
5 in the hundredths place

_____

14. 7 in the tenths place
0 in the hundredths place
9 in the tens place
4 in the ones place

_____

15. 9 in the ones place
4 in the hundredths place
3 in the tenths place
8 in the thousandths place

_____

16. 2 in the tens place
5 in the tenths place
9 in the ones place
6 in the hundredths place

_____

17. 0 in the tenths place
4 in the thousandths place
7 in the hundredths place
2 in the ones place

_____

Solve each problem. You can draw pictures or use counters.

18. Ginny bought 4 boxes of markers. Each box has 8 markers.

How many markers did Ginny buy? _____

19. John has 18 roses. He puts 9 roses in each vase.

How many vases does he fill? _____

How many roses are left over? _____

20. Lia shared 22 cookies equally among 6 friends.

How many cookies did each friend get? _____

How many cookies were left over? _____

Use with or after Lesson 6·12.

Practice Set 56

Draw a square array for each square number. Then write the multiplication fact for each square number.

Example   25   • • • • • $5 \times 5 = 25$
                    • • • • •
                    • • • • •
                    • • • • •
                    • • • • •

1. 36   **2.** 16   **3.** 4   **4.** 49

Find the missing numbers. You can use counters or draw pictures.

5. 26 crackers
6 children share equally

_____ crackers per child

_____ crackers left over

6. 36 pictures
9 pictures per page

_____ filled pages

_____ pages left over

7. 24 girls
4 girls per tent

_____ filled tents

_____ girls left over

8. 18 sheets of paper
4 children share equally

_____ sheets per child

_____ sheets left over

Find the corresponding letter on the centimeter ruler for each of the metric measures.

Example   2.5 centimeters is Point *A*.

A G   D   B   E   F   C

```
0  1  2  3  4  5  6  7  8  9  10  11  12  13  14  15
cm
```

9. 60 millimeters _____   **10.** 1 decimeter _____   **11.** 0.04 meter _____

12. 0.7 decimeter _____   **13.** 85 millimeters _____   **14.** 3 centimeters _____

Practice Set 56 continued

Write the number that is 10 more.

15. 14 _____ **16.** 30 _____ **17.** 539 _____ **18.** 4,258 _____ **19.** 7,904 _____

Write the number that is 100 more.

20. 8 _____ **21.** 27 _____ **22.** 973 _____ **23.** 2,918 _____ **24.** 8,715 _____

Write the number that is 1,000 more.

25. 7 _____ **26.** 254 _____ **27.** 5,791 _____ **28.** 9,493 _____ **29.** 12,463 _____

Write the number that is 10 less.

30. 19 _____ **31.** 142 _____ **32.** 1,014 _____ **33.** 7,420 _____ **34.** 4,615 _____

Write the number that is 100 less.

35. 156 _____ **36.** 433 _____ **37.** 5,212 _____ **38.** 1,082 _____ **39.** 12,617 _____

Write the number that is 1,000 less.

40. 1,092 _____ **41.** 7,214 _____ **42.** 5,131 _____ **43.** 10,673 _____ **44.** 22,194 _____

Find each answer using mental math.

45. $70 - 20 =$ _____ **46.** $300 + 600 + 500 =$ _____ **47.** $800 - 600 =$ _____

48. $1,200 - 500 =$ _____ **49.** $6,300 + 800 =$ _____ **50.** $12,000 + 500 =$ _____

51. **Writing/Reasoning** Sherri had $1,200 in her savings account. Then she took out $700. The next week she put $900 into her account. The following week, Sherri took out $300. How much is in her account now? Explain your answer.

_____

_____

Practice Set 57

Write the missing number for each Fact Triangle.
Then write the family of facts for that triangle.

1.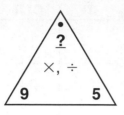

Missing number: _____

Fact family:

_____

_____

_____

2.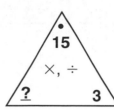

Missing number: _____

Fact family:

_____

_____

_____

3.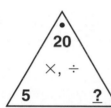

Missing number: _____

Fact family:

_____

_____

_____

4.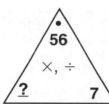

Missing number: _____

Fact family:

_____

_____

_____

5.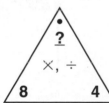

Missing number: _____

Fact family:

_____

_____

_____

6.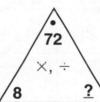

Missing number: _____

Fact family:

_____

_____

_____

 Multiply. Remember to practice and memorize your multiplication facts.

7. $2 \times 4 =$ _____   **8.** $6 \times 8 =$ _____   **9.** $5 \times 9 =$ _____   **10.** $3 \times 7 =$ _____

11. $9 \times 7 =$ _____   **12.** $7 \times 8 =$ _____   **13.** $9 \times 8 =$ _____   **14.** $7 \times 6 =$ _____

15. $9 \times 9 =$ _____   **16.** $4 \times 8 =$ _____   **17.** $8 \times 4 =$ _____   **18.** $6 \times 6 =$ _____

Practice Set 57 continued

Write the missing rule and the missing numbers.

Example

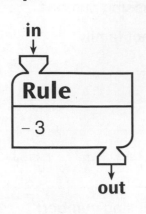

Rule − 3

in	out
5	2
11	8
17	14
23	20
19	16

19.

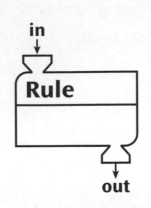

Rule

in	out
4	16
10	22
	24
12	
20	

20.

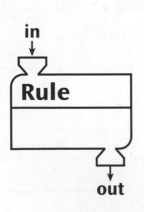

Rule

in	out
2	6
3	9
	12
6	
8	

21.

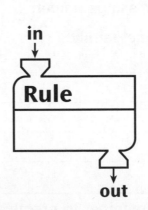

Rule

in	out
8	4
10	5
	3
14	
18	

Use with or after Lesson 7·2.

Practice Set 58

 Multiply. Remember to practice and memorize your multiplication facts.

1. $6 \times 6 =$ _____   **2.** $8 \times 3 =$ _____   **3.** $9 \times 6 =$ _____   **4.** $9 \times 9 =$ _____

5. $7 \times 6 =$ _____   **6.** $8 \times 8 =$ _____   **7.** $8 \times 9 =$ _____   **8.** $8 \times 4 =$ _____

9. $5 \times 8 =$ _____   **10.** $9 \times 4 =$ _____   **11.** $8 \times 6 =$ _____   **12.** $7 \times 7 =$ _____

13. $8 \times 6 =$ _____   **14.** $7 \times 8 =$ _____   **15.** $9 \times 5 =$ _____   **16.** $7 \times 9 =$ _____

Denise invented a game using this gameboard. Answer each question below.

17. How many rows are on Denise's gameboard? _____

How many squares are in each row? _____

18. Write a number model to show the total number of squares

on Denise's gameboard. _____

19. How many of the squares on the gameboard are black?

_____

How many are white? _____

20. If 2 markers can be placed on each square of the gameboard,

how many markers can the gameboard hold? _____

Practice Set 59

Write a number model. Then solve.

> **Example**   Jerry picked **50** apples. He ate **2** of them. Then he
> divided the rest of the apples equally into **8** baskets.
> How many apples did he put in each basket?
> **Number model: (50 − 2) ÷ 8 = 6**

1. Karen made 3 clay pots Monday and 4 clay pots Tuesday.
By the end of the week, she had made 17 pots. How many
pots did she make between Wednesday and Friday?

_____

2. Keneisha had 18 stickers. She put 2 of them on her notebook
and the rest on her folders. She put 4 of them on each of her
folders. How many folders did Keneisha have?

_____

3. Franklin has 82 seashells. He wants to have 100. His friend
Mario gave him 7. How many more shells does Franklin need?

_____

4. Tim needs 24 cupcakes for his birthday party. His mother made 12.
Tim has 4 friends who said they will bring the rest. How many cupcakes
should each friend bring if all 4 friends bring equal amounts?

_____

Find each answer.

5. For a picnic, Sharon brought 3 cookies for each
of 4 people. How many cookies did she bring in all? _____

6. Tom bought 2 packages of postcards. Each package
contained 5 postcards. How many postcards did he buy in all? _____

Practice Set 59 continued

Write each number.

> **Example**
> one million, four hundred ten thousand, five hundred three   **1,410,503**

7. three million, nine hundred fifty-four thousand, six hundred twenty-nine _____

8. two million, thirty-nine thousand, four hundred ninety-eight _____

9. nine hundred forty-one thousand, eight hundred five _____

10. seven million, three thousand, two hundred eighty _____

11. nine million, eight hundred two _____

12. six million, nine thousand, ten _____

Write the multiplication and division fact family for each group of numbers.

13. 45, 9, 5

14. 20, 4, 80

15. 6, 40, 240

_____   _____   _____

_____   _____   _____

_____   _____   _____

_____   _____   _____

Basketball players can shoot baskets worth 3 points, 2 points, and 1 point. Find three different ways that a player can score 10 points.

16. _____   **17.** _____   **18.** _____

19. Make a number model to show the points a basketball player made by getting two 3-point baskets, one 2-point basket, and two 1-point baskets.

_____

20. **Writing/Reasoning** Write a story that can be solved by this number sentence: $10 - (3 + 2) = 5$.

_____

_____

Practice Set 60

SRB
52 53

Answer each question.

1. How much are 7 [80s]? _____

2. How much are 4 [600s]? _____

3. Which number multiplied by 3 equals 90? _____

4. Which number multiplied by 4 equals 1,600? _____

Solve each problem.

5. Sharon bought 3 packages of hair bows. There are 5 bows in each package. How many bows did Sharon buy?

_____

6. Joe got 4 packages of stickers as a gift. Each package holds 6 stickers. How many stickers did Joe get?

_____

7. A sheet of stamps has 6 rows. Each row has 3 stamps. How many stamps are on a sheet?

_____

8. Each box of crackers holds 300 crackers. You have no boxes of crackers. How many crackers do you have?

_____

9. Each row of buttons has 6 buttons. You have 1 row of buttons. How many buttons do you have?

_____

10. 5 cakes are each cut into 6 pieces. How many pieces of cake are there?

_____

FACTS PRACTICE

Multiply. Remember to practice and memorize your multiplication facts.

11. $2 \times 7 =$ _____

12. $4 \times 3 =$ _____

13. $9 \times 8 =$ _____

14. $5 \times 2 =$ _____

15. $7 \times 6 =$ _____

16. $6 \times 8 =$ _____

17. $8 \times 8 =$ _____

18. $3 \times 9 =$ _____

19. $3 \times 3 =$ _____

20. $9 \times 9 =$ _____

21. $7 \times 3 =$ _____

22. $4 \times 9 =$ _____

Use with or after Lesson 7•6.

Practice Set **60** continued

Measure each line segment to the nearest centimeter. Then tell whether the line segments are *parallel* or *intersecting.*

Example

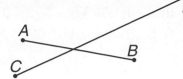

$\overline{AB}$ is 3 cm long.

$\overline{CD}$ is 5 cm long.

$\overline{AB}$ and $\overline{CD}$ are intersecting.

23.

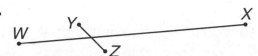

_____

_____

_____

24.

_____

_____

_____

_____

25.

_____

_____

_____

Finish each symmetrical shape according to the line of symmetry.

Example

26.    **27.** 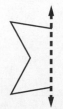   **28.**    **29.**

Practice Set 61

SRB
52 53
190–192

Use estimation to solve each of these problems:

1. Laura has $45.00. Does she have enough to buy 2 skirts that each cost $14.50 and a blouse that costs $17.00?

_____

2. Dennis has $80.00. Does he have enough to buy gifts for his family that cost $23.50, $18.90, $15.95, and $17.50?

_____

3. Linda earned $33.00 in January and $22.00 in February. She paid her sister back $9.00 that she owed her. Does Linda have enough money left to pay for a weekend trip that costs $48.00?

_____

4. Dave earned $24.00 in July by cutting lawns. He earned $29.00 in August and $17.00 in September. Did Dave earn enough money to buy 2 games that cost $33.99 each?

_____

5. **Writing/Reasoning** Jill, JoAnn, and Jackie want go to an amusement park. The admission fee for each girl will be $22.95. Lunch for each girl will cost $5.50. Each girl received $30.00 for her birthday. Do they have enough money to go to the amusement park? Explain your answer.

_____

_____

Find each missing number.

6. $4 = 2 \times$ _____

7. _____ $= 8 \times 0$

8. _____ $\times 5 = 25$

9. $8 \div 1 =$ _____

10. $4 \times$ _____ $= 4$

11. $63 =$ _____ $\times 7$

12. $9 =$ _____ $\div 1$

13. _____ $= 3 \times 6$

14. $27 =$ _____ $\times 9$

15. $4 \times 7 =$ _____

16. $3 \times$ _____ $= 12$

17. _____ $\times 10 = 30$

18. _____ $= 50 \times 1$

19. $1 \div$ _____ $= 1$

20. _____ $\times 6 = 18$

21. $7 =$ _____ $\div 1$

22. _____ $\times 6 = 36$

23. $1 \times$ _____ $= 0$

Practice Set 62

SRB
96–101

 **Multiply.**

1. 7 × 90 = _____   **2.** 40 × 40 = _____   **3.** 10 × 280 = _____   **4.** 5 × 50 = _____

5. 70 × 6 = _____   **6.** 400 × 8 = _____   **7.** 69 × 10 = _____   **8.** 28 × 10 = _____

9. 100 × 30 = _____   **10.** 70 × 30 = _____   **11.** 500 × 2 = _____   **12.** 8 × 300 = _____

Draw the following figures:

13. 2 parallel lines

14. 2 intersecting line segments

15. 2 rays that form a right angle

16. an angle that shows a half-turn

Find the median for each set of numbers below.

17. 27 50 42 18 42

18. 36 9 17 24 28 15 14

_____

_____

19. 82 75 79 81

20. 15 30 19 28 34 11

_____

_____

21. 62 28 55 49 38

22. 83 34 68 58 68 97

_____

_____

Practice Set 62 continued

Show each amount using the fewest coins and bills possible.

Use $1 s, Q s, D s, N s, and P s.

Example $1.86

$1   Q   Q   Q   D   P

23. 79¢   **24.** $0.93   **25.** $1.52

26. 49¢   **27.** $0.65   **28.** $2.19

29. $0.98   **30.** $3.84   **31.** 59¢

Solve each problem.

32. Dave caught a fish that was 26 inches long. Kim caught a fish that was 42 inches long. How much longer was Kim's fish?

33. The lowest temperature in Denver one year was 8°F. The highest temperature during that same year was 96°F. What was the difference between the two temperatures?

34. In the morning, the temperature was 24°F. By 2:00 in the afternoon, the temperature was 47°F. How much had the temperature risen?

35. Larry planted a bush that was 38 centimeters tall. Three months later the bush was 51 centimeters tall. How much had the bush grown?

 Add or subtract. Remember to practice and memorize your basic facts.

36.
$$\begin{array}{r} 532 \\ -\ 128 \\ \hline \end{array}$$

37.
$$\begin{array}{r} 115 \\ +\ 247 \\ \hline \end{array}$$

38.
$$\begin{array}{r} 879 \\ -\ 254 \\ \hline \end{array}$$

39.
$$\begin{array}{r} 647 \\ +\ 131 \\ \hline \end{array}$$

Practice Set 63

Use the following information to answer the questions below:

A school cafeteria can spend $1.50 on each student per lunch. One pizza costs the school $1.50. One hot dog, however, costs the school only $0.50.

1. How many hot dogs can replace 1 pizza? _____

2. How many hot dogs can replace 2 pizzas? _____

3. How many hot dogs can replace 50 pizzas? _____

4. How many hot dogs can replace 400 pizzas? _____

Write the number that has ...

5. 1 hundred-thousand
4 tens
5 ten-thousands
7 ones
9 thousands
3 hundreds

_____

6. 7 ten-thousands
2 ones
9 hundreds
0 thousands
3 hundred-thousands
9 tens

_____

7. 4 hundreds
8 ones
5 hundred-thousands
2 tens
9 thousands
0 ten-thousands

_____

8. 6 thousands
7 ten-thousands
6 ones
8 hundreds
4 tens
5 hundred-thousands

_____

 Multiply. Remember to practice and memorize your multiplication facts.

9. $5 \times 4 =$ _____   **10.** $2 \times 1 =$ _____   **11.** $8 \times 8 =$ _____   **12.** $3 \times 6 =$ _____

13. $7 \times 9 =$ _____   **14.** $4 \times 7 =$ _____   **15.** $5 \times 7 =$ _____   **16.** $9 \times 2 =$ _____

17. $4 \times 9 =$ _____   **18.** $9 \times 8 =$ _____   **19.** $6 \times 5 =$ _____   **20.** $10 \times 4 =$ _____

Practice Set 63 *continued*

Write the letter of the square or rectangle that matches
each description below.

21. a rectangle with a perimeter of 18 units _____

22. a square with an area of 25 square units _____

23. a rectangle with an area of 10 square units _____

24. a rectangle with a perimeter of 10 units _____

25. a square with a perimeter of 20 units _____

26. a square that has the same number
for its perimeter and its area _____

27. a rectangle that has an area of 14 square units _____

28. a rectangle that has a perimeter of 14 units _____

A.

B.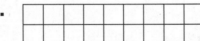

C.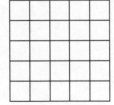

D.

E.

F.

Use with or after Lesson 7•9.

Practice Set 64

Write the fraction for the shaded part of each picture.

Example  $\dfrac{2}{5}$

1. 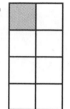 _____

2.   _____

3. 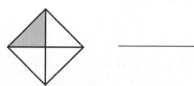 _____

4. _____

5. 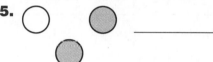 _____

6.  _____

7. 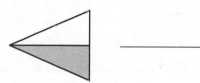 _____

8. 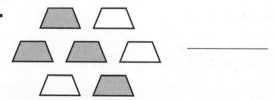 _____

 Multiply. Remember to practice and memorize your multiplication facts.

9. $3 \times 6 =$ _____ **10.** $6 \times 5 =$ _____ **11.** $9 \times 7 =$ _____ **12.** $8 \times 3 =$ _____

13. $2 \times 1 =$ _____ **14.** $0 \times 8 =$ _____ **15.** $4 \times 9 =$ _____ **16.** $9 \times 2 =$ _____

Practice Set 65

Teresa has a blank six-sided die. She writes the number 1 on one side. Predict what will happen in the following situations.

1. Teresa rolls the die once. Is the number 1 *likely, unlikely,* or *certain* to come up?

_____

2. Teresa writes the number 2 on another side of the die. Is it now *likely, unlikely,* or *certain* that a number will come up on one roll?

_____

3. **Writing/Reasoning** Teresa writes the number 3 on a third side of the die. About how many times will a number come up if she rolls the die 10 times? Explain your answer.

_____

_____

_____

_____

Solve.

4. $4.56 + $2.31 = _____   **5.** $10.00 − $7.76 = _____

6. $1.30 + $6.50 = _____   **7.** $9.70 − $3.43 = _____

Fill in the missing numbers.

8. $0.25, _____, _____, _____, $1.25, $1.50, _____, _____

9. $6.00, _____, _____, _____, $4.00, $3.50, _____, _____

10. $1.80, _____, _____, $2.10, $2.20, _____, _____, _____

Practice Set 66

Find the missing numbers on each number line.

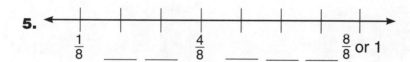

1. ← | — $\frac{1}{3}$ — | — — — | — $\frac{3}{3}$ or 1 — →

2. ← — — | — — — | — $\frac{2}{2}$ or 1 — →
   with — under the middle mark

3. ← | $\frac{1}{4}$ — — | — — | — — | — $\frac{4}{4}$ or 1 →

4. ← | — — | — — | $\frac{3}{6}$ — | — — | — — | $\frac{6}{6}$ or 1 →

5. ← | $\frac{1}{8}$ — | — — | — $\frac{4}{8}$ — | — — | — — | — — | $\frac{8}{8}$ or 1 →

Find each missing number. You can use the number lines above to help you.

> **Example**
>
> $1 = \dfrac{6}{6}$

6. $\dfrac{1}{2} = \dfrac{}{8}$

7. $\dfrac{4}{4} = \dfrac{}{3}$

8. $\dfrac{}{6} = \dfrac{2}{3}$

9. $\dfrac{}{8} = 1$

10. $\dfrac{6}{8} = \dfrac{}{4}$

11. $\dfrac{1}{} = \dfrac{2}{6}$

12. $\dfrac{3}{3} = \dfrac{8}{}$

13. $\dfrac{}{8} = \dfrac{1}{4}$

14. $\dfrac{}{2} = \dfrac{3}{} = \dfrac{4}{8} = \dfrac{}{4}$

15. $\dfrac{}{7} = 0$

Use with or after Lesson 8·3.

109

Practice Set 67

Write as many numbers as you can for the fractional parts
shown in each picture.

Example

$\frac{10}{10}$ or 1 $\frac{5}{10}$ or $\frac{1}{2}$

1.

2.

$\frac{8}{8}$ or 1

_____   _____

3.

4.

$\frac{6}{6}$ or 1

_____   _____

5.

6.

$\frac{5}{5}$ or 1

_____   _____

7.

8.

$\frac{4}{4}$ or 1

_____   _____

Practice Set 67 continued

Measure each object to the nearest half-inch.

Example

about 2$\frac{1}{2}$ inches

9.

_____

10.

_____

11.

_____

Multiply. Remember to practice and memorize your multiplication facts.

12. $1 \times 8 =$ _____

13. $0 \times 5 =$ _____

14. $15 \times 1 =$ _____

15. $24 \times 0 =$ _____

16. $1 \times 37 =$ _____

17. $145 \times 0 =$ _____

Practice Set 68

Write <, >, or =.

Example   This is ONE: $\frac{3}{6}$ _____ $\frac{2}{6}$

$$\frac{3}{6} > \frac{2}{6}$$

1. This is ONE:

$\frac{1}{2}$ _____ $\frac{2}{4}$ $\frac{3}{4}$ _____ $\frac{1}{4}$

2. This is ONE:

$\frac{5}{8}$ _____ $\frac{3}{4}$ $\frac{1}{2}$ _____ $\frac{4}{8}$

3. This is ONE:

$\frac{1}{2}$ _____ $\frac{3}{4}$ $\frac{1}{3}$ _____ $\frac{1}{6}$

4. This is ONE:

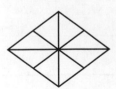

$\frac{3}{8}$ _____ $\frac{1}{4}$ $\frac{2}{8}$ _____ $\frac{1}{4}$

         Use with or after Lesson 8·5.

Practice Set 68 *continued*

Write the missing numbers in the tables.

5.

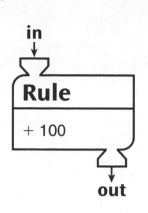

in	out
17	117
106	206
314	
1,462	
5,067	

6.

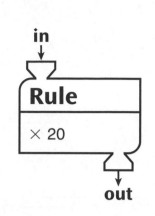

in	out
6	120
8	
20	400
30	
90	

7.

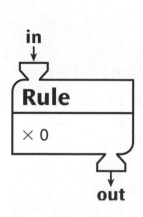

in	out
5	0
19	
108	
2,416	
9,999	

8.

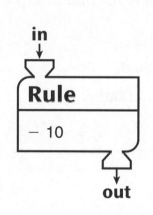

in	out
15	5
254	
306	
4,000	
8,406	

Fill in the missing numbers.

9. 8, _____, 12, 14, _____, _____, _____, 22

10. _____, 10, 15, 20, _____, _____, _____, _____

11. 50, _____, _____, 80, 90, _____, _____, 120

Practice Set 69

Write a fraction and a mixed number to match each picture.

1.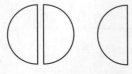

   _____   _____

2.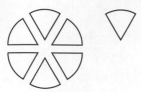

   _____   _____

3.

   _____   _____

Find the missing number for each Fact Triangle.
Then write the fact family for that triangle.

4.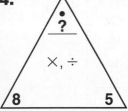

   Missing number: ___

   Fact family:

   _____

   _____

   _____

5.

   Missing number: ___

   Fact family:

   _____

   _____

   _____

6.

   Missing number: ___

   Fact family:

   _____

   _____

7.

   Missing number: ___

   Fact family:

   _____

   _____

Use with or after Lesson 8•6.

Practice Set 69 *continued*

 Complete the number models.

8. $(4 \times 3) + 12 =$ _____

9. $40 = (24 + 36) -$ _____

Use the numbers in each addition sentence below to write another addition sentence and two subtraction sentences.

> **Example**
> $19 + 6 = 25$
> $6 + 19 = 25$
> $25 - 6 = 19$
> $25 - 19 = 6$

10. $8 + 6 = 14$

_____

_____

_____

11. $5 + 9 = 14$

_____

_____

_____

12. $7 + 34 = 41$

_____

_____

_____

13. $10 + 7 = 17$

_____

_____

_____

14. $56 + 8 = 64$

_____

_____

_____

15. $40 + 30 = 70$

_____

_____

_____

16. $100 + 4 = 104$

_____

_____

_____

17. $93 + 6 = 99$

_____

_____

_____

18. $200 + 500 = 700$

_____

_____

_____

Find an equal amount of money in the second list.
Write the letter that identifies that amount.

19. $\frac{1}{100}$ dollar _____

20. $\frac{1}{5}$ quarter _____

21. quarter _____

22. $\frac{1}{2}$ dollar _____

23. 10¢ _____

24. $0.75 _____

A. $0.05

B. $\frac{1}{10}$ dollar

C. penny

D. 25¢

E. 50¢

F. $\frac{3}{4}$ dollar

Use with or after Lesson 8·6.

Practice Set 70

Solve each problem.

1. Sharon brought 12 apples to the picnic. After the picnic, 2 apples were left. What fraction of the apples were eaten?

   _____

2. Dave spent 5 days at camp. What fraction of a week did Dave spend at camp?

   _____

3. Dorothy bought 10 yards of ribbon. She used 2 yards to wrap packages. What fraction of the ribbon did she use?

   _____

4. Glenda had $15. She spent $9 on a book. What fraction of her money did Glenda spend on the book?

   _____

5. For a party, a huge sandwich was cut into 25 pieces. After the party, 5 pieces were left. What fraction of the sandwich was eaten? What fraction of the sandwich was not eaten?

   _____

   _____

6. A vase of flowers has 6 red roses, 6 yellow roses, and 12 white roses. What fraction of the roses are yellow? What fraction of the flowers are white?

   _____

   _____

Make your own name-collection box. Include +, −, ×, and ÷.
Include at least 8 different names for each number.

7.

45

8.

30

Use with or after Lesson 8•7.

Practice Set 71

Solve each problem.

1. Marsha is reading a 250-page book. She has
   read 50 pages. What fraction of the book has she read?

   _____

2. Nathan wants to buy a gift for his brother. The gift
   costs $8.00. Nathan has $12.00. What fraction of
   his money must he spend to buy the gift?

   _____

3. Katie and her family are driving to an amusement park.
   The park is 300 miles from their home. They have driven
   $\frac{1}{3}$ of the distance. How far are they from the park?

   _____

4. **Writing/Reasoning** Hannah and Katrina are watching a
   90-minute movie. After 45 minutes, Hannah says it is $\frac{1}{2}$ over.
   Katrina insists it is $\frac{2}{3}$ over. Who is correct? Explain your answer.

   _____

   _____

   _____

5.

   How many fifths? _____ fifths   Color 6 fifths.

   Write the fraction. _____   Write the mixed number. _____

6.

   How many eighths? _____ eighths   Color 10 eighths.

   Write the fraction. _____   Write the mixed number. _____

Practice Set 72

Solve each problem.

1. a. 200 [400s] _____   **b.** 200 × 400 = _____

2. a. 300 [500s] _____   **b.** 300 × 500 = _____

3. How many 500s are in 2,000? _____

4. How many 300s are in 3,000? _____

5. How many 200s are in 2,000? _____

FACTS PRACTICE   **Solve each problem. Circle the square products.**

6. 7 × 4 = _____   **7.** 4 × 4 = _____   **8.** 6 × 6 = _____

9. 9 × 9 = _____   **10.** 8 × 8 = _____   **11.** 7 × 8 = _____

12. 6 × 9 = _____   **13.** 7 × 7 = _____   **14.** 9 × 8 = _____

15. 5 × 5 = _____   **16.** 8 × 6 = _____   **17.** 7 × 9 = _____

Write a mixed number for each fraction. Draw pictures to help you.

Example $\frac{6}{4}$ $1\frac{2}{4}$

18. $\frac{3}{2}$ _____   **19.** $\frac{6}{5}$ _____   **20.** $\frac{7}{3}$ _____

118

Practice Set 73

Solve the following problems mentally.

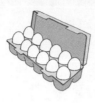 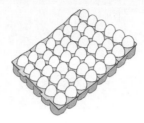

Carton   **Tray**

1. How many eggs are in 2 cartons? _____

2. How many eggs are in 1 tray? _____

3. How many eggs are in half a tray? _____

4. How many eggs are in 3 cartons? _____

5. How many eggs are in half a carton? _____

6. Which is more, 4 cartons or 1 tray? _____

Find each missing number.

Example	double 2 gloves = **4 gloves**
	triple 2 gloves = **6 gloves**
	quadruple 2 gloves = **8 gloves**
	5 times 2 gloves = **10 gloves**
	10 times 2 gloves = **20 gloves**

7. 3¢

double 3¢ = _____ ¢

triple 3¢ = _____ ¢

quadruple 3¢ = _____ ¢

5 times 3¢ = _____ ¢

10 times 3¢ = _____ ¢

8. 4 inches

double 4 in. = _____ in.

triple 4 in. = _____ in.

quadruple 4 in. = _____ in.

5 times 4 in. = _____ in.

10 times 4 in. = _____ in.

Use with or after Lesson 9·2.

Practice Set 73 continued

Write the fraction for the shaded part of the picture
in two different ways.

Example 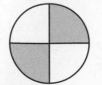 $\frac{1}{2}$ or $\frac{2}{4}$

9.
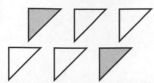

_____

10.

_____

11.

_____

12.

_____

13.
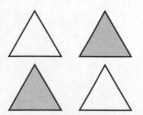

_____

14.

_____

 Multiply. Remember to practice and memorize
your multiplication facts.

15. $2 \times 5 =$ _____

16. $8 \times 6 =$ _____

17. $4 \times 7 =$ _____

18. $9 \times 9 =$ _____

19. $3 \times 4 =$ _____

20. $7 \times 5 =$ _____

Use with or after Lesson 9·2.

Practice Set 74

Solve.

1. Veronica studies one hour after school each day. If she has two hours after school each day to study and volunteer, what fraction of the time does she volunteer? _____

2. Mrs. Jones cut a pan of brownies into 12 pieces. Her family ate 6 pieces at dinner. Ethan ate $\frac{1}{3}$ of the pieces that the family ate. What fraction of the original 12 brownies did he eat?

   _____

Measure each line segment to the nearest $\frac{1}{4}$ inch.

3.

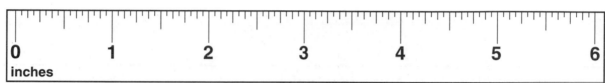

_____

4.

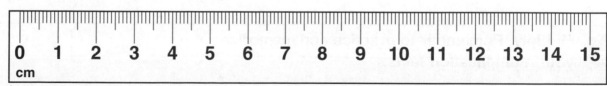

_____

Measure each line segment to the nearest $\frac{1}{2}$ centimeter.

5.

6.

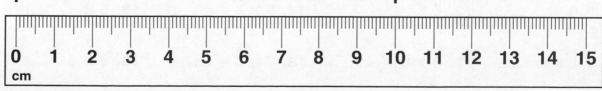

_____

Practice Set 75

SRB
52 53

Multiply. Then check your answers using a calculator.

Example	29
	× 3
3 [20s]	60
3 [9s]	27
60 + 27	**87**

1. 51
    × 6

2. 84
    × 5

3. 17
    × 8

4. 206
    × 9

5. 419
    × 4

**For each problem below, write a number model.
Then find the missing numbers.**

6. Donna puts 6 pears in each bag. She has 32 pears. How many bags does she fill?

_____ ÷ _____ → _____ R _____

Donna fills _____ bags.

_____ pears are left over.

7. 21 signs are shared equally by 4 classrooms. How many signs does each classroom get?

_____ ÷ _____ → _____ R _____

Each classroom gets _____ signs.

_____ signs are left over.

FACTS PRACTICE

Multiply. Remember to practice and memorize your multiplication facts.

8. $7 \times 8 =$ _____   **9.** $1 \times 9 =$ _____   **10.** $3 \times 5 =$ _____   **11.** $4 \times 4 =$ _____

12. $2 \times 6 =$ _____   **13.** $6 \times 7 =$ _____   **14.** $9 \times 3 =$ _____   **15.** $5 \times 8 =$ _____

16. $6 \times 8 =$ _____   **17.** $9 \times 8 =$ _____   **18.** $7 \times 6 =$ _____   **19.** $8 \times 8 =$ _____

Use with or after Lesson 9·4.

Practice Set 76

Use the grocery store sign to solve each number story.

```
————  Special  ————
Peaches      $0.50  each
Apples       $0.39  each
Pears        $0.61  each
            (No tax)
```

1. Sandra has $2.00. Can she buy 5 apples? _____

How much money will she have left? _____

2. How much money does Kenny need to buy 4 pears? _____

3. How much does it cost to buy 6 pears and 3 apples? _____

FACTS PRACTICE **Multiply. Circle each square product.**

4. $7 \times 9 -$ _____ **5.** $8 \times 8 -$ _____ **6.** $6 \times 7 =$ _____

7. $4 \times 8 =$ _____ **8.** $5 \times 5 =$ _____ **9.** $9 \times 9 =$ _____

10. $6 \times 6 =$ _____ **11.** $7 \times 7 =$ _____ **12.** $8 \times 9 =$ _____

13. $8 \times 7 =$ _____ **14.** $6 \times 8 =$ _____ **15.** $6 \times 6 =$ _____

Change each mixed number to a fraction. Draw pictures to help you.

Example $1\frac{2}{3}$ ⬤⬤ $\frac{5}{3}$

16. $2\frac{1}{2}$ _____ **17.** $3\frac{3}{4}$ _____ **18.** $1\frac{3}{6}$ _____

Practice Set **76** *continued*

SRB
254–260

Solve each problem.

19. Each red fox weighs 19 pounds. How much do 7 red foxes weigh?

_____

20. Ben spent 25 minutes walking to school. What fraction of an hour is this? (*Hint:* 1 hour = 60 minutes.)

_____

21. Betty has 25 stickers. She wants to share them equally among 3 friends. How many stickers will each friend get? How many stickers will be left over?

_____

22. Ellen had $30.00. She spent $14.00 shopping. What fraction of her money did she spend? What fraction of her money did she NOT spend?

_____

23. Sam's birthday cake was cut into 16 pieces. After his party, 3 pieces were left. What fraction of his cake was left? What fraction of his cake was eaten?

_____

24. Ruth wants to buy 4 computer games that each cost $29.50. About how much money does Ruth need in order to buy all 4 computer games?

_____

25. Jan wants to get a haircut and buy shampoo. Jan has $25.00. Does she have enough money for a haircut that costs $16.00 and 2 bottles of shampoo that cost $4.25 each?

_____

26. A quilt has 5 yellow squares, 10 blue squares, and 10 green squares. What fraction of the squares are blue? What fraction of the squares are yellow?

_____

Use with or after Lesson 9·5.

Practice Set 77

Read the information below. Then answer each question.

*Ted has 36 flowers. He wants to put the flowers into vases.
He wants each vase to have the same number of flowers—
without any flowers being left over.*

1. Can he put the flowers in 1 vase? 2 vases? 3 vases?

If so, how many flowers go in each vase? _____

2. Can he put the flowers in 4 vases? 5 vases? 6 vases?

If so, how many flowers go in each vase? _____

3. Can he put the flowers in 7 vases? 8 vases? 9 vases?

If so, how many flowers go in each vase? _____

4. Can he put the flowers in 10 vases? 11 vases? 12 vases?

If so, how many flowers go in each vase? _____

The factors of 36 are the numbers that can be multiplied by whole numbers to
get 36, or the numbers that 36 can be divided by without having remainders.

5. Name the factors of 36. _____

For each number below, give the value of each digit.

Example 48.613   4 tens
                        8 ones
                        6 tenths
                        1 hundredth
                        3 thousandths

6. 295.6 _____   **7.** 30.48 _____   **8.** 10.925 _____

_____   _____   _____

_____   _____   _____

_____

Practice Set 77 continued

Write the number family for each Fact Triangle.

Example

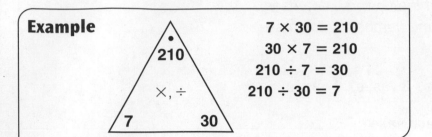

$7 \times 30 = 210$

$30 \times 7 = 210$

$210 \div 7 = 30$

$210 \div 30 = 7$

9.

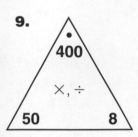

_____

_____

_____

10.

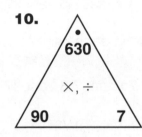

_____

_____

_____

11.

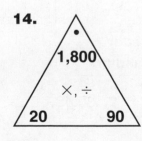

_____

_____

_____

12.

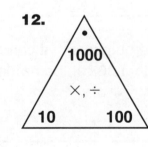

_____

_____

_____

13.

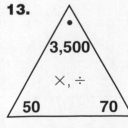

_____

_____

_____

14.

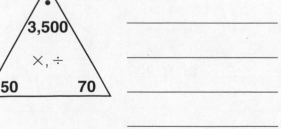

_____

_____

_____

126

Practice Set 78

SRB
10–12
52 53

Solve each problem.

1. $56 ÷ 8 = $ _____   **2.** $81 ÷ 9 = $ _____   **3.** $54 ÷ 6 = $ _____

4. $150 ÷ 6 = $ _____   **5.** $120 ÷ 8 = $ _____   **6.** $140 ÷ 7 = $ _____

7. $122 ÷ 4 = $ _____   **8.** $85 ÷ 5 = $ _____   **9.** $490 ÷ 7 = $ _____

Write the missing fractions or mixed numbers.

10.

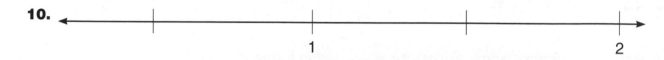

11.

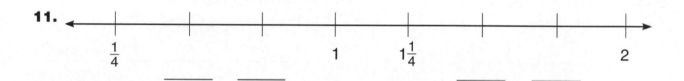

12.

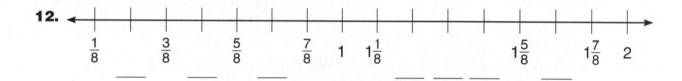

13.

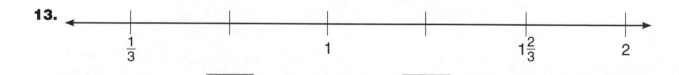

14.

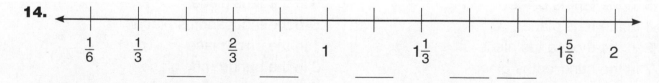

Write the mixed number as a fraction.

15. $1\frac{1}{2}$ _____   **16.** $2\frac{3}{5}$ _____   **17.** $1\frac{2}{3}$ _____   **18.** $3\frac{1}{2}$ _____

19. $2\frac{3}{4}$ _____   **20.** $1\frac{1}{4}$ _____   **21.** $3\frac{1}{9}$ _____   **22.** $2\frac{1}{12}$ _____

Practice Set 79

Solve each problem.

1. Ken wants to put 6 ounces of water in each glass. How many glasses can he fill with 42 ounces of water? How many ounces of water will be left over?

_____

2. Lynn wants to cut a 50-inch piece of string into pieces that are each 8 inches long. How many 8-inch pieces can she cut? How many inches of string will be left over?

_____

Write the number that has ...

3. 7 in the tens place
1 in the thousands place
4 in the tenths place
2 in the hundreds place
6 in the ones place

_____

4. 4 in the hundredths place
1 in the tens place
5 in the ones place
7 in the tenths place
9 in the hundreds place

_____

5. 0 in the tenths place
9 in the thousandths place
2 in the ones place
5 in the tens place
8 in the hundredths place

_____

6. 0 in the hundredths place
6 in the ones place
1 in the tens place
9 in the thousandths place
3 in the tenths place

_____

7. 3 in the hundreds place
9 in the ones place
8 in the tenths place
4 in the tens place
6 in the thousands place
7 in the hundredths place

_____

8. 4 in the tenths place
2 in the thousands place
0 in the tens place
6 in the thousandths place
1 in the ones place
0 in the hundredths place
4 in the hundreds place

_____

Use with or after Lesson 9•8.

Practice Set 80

Use lattice multiplication to solve each problem.

1. 8 × 49 = _____

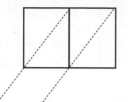

2. 7 × 359 = _____

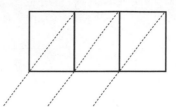

3. 6 × 314 = _____

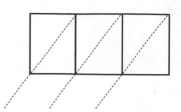

4. 9 × 68 = _____

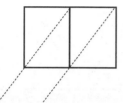

5. 5 × 456 = _____

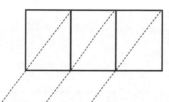

6. 7 × 834 = _____

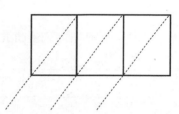

Write a multiplication fact to find the area of each square.

Example   3 × 3 = 9
Area = 9 square units

7.

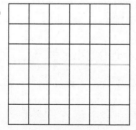

8.

9.

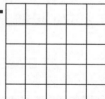

_____ × _____ = _____   _____ × _____ = _____   _____ × _____ = _____

Area = _____   Area = _____   Area = _____

Practice Set 81

Write each number in base-10 block shorthand. Use as few blocks as possible.

Example 225

1. 124

_____

2. 85

_____

3. 213

_____

4. 261

_____

5. 304

_____

Solve.

6. $8 \times 46 =$ _____

7. $7 \times 53 =$ _____

8. $4 \times 62 =$ _____

9. $3 \times 154 =$ _____

10. $6 \times 432 =$ _____

11. $9 \times 610 =$ _____

12. $5 \times 731 =$ _____

13. $7 \times 208 =$ _____

14. $4 \times 836 =$ _____

Tell what the underlined digit stands for in each number.

15. 8<u>6</u>0,345 _____

16. <u>4</u>,900 _____

17. 590,3<u>4</u>0 _____

Answer the following questions about the triangle.

18. What type of triangle is $\triangle DEF$? _____

19. What parts of $\triangle DEF$ are equal? _____

20. What is the perimeter of $\triangle DEF$? _____

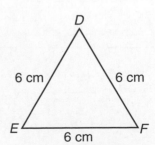

Use with or after Lesson 9•10.

Practice Set 82

Multiply using the partial-products method. Then use a calculator to check each answer.

Example	78
	× 43
(40 × 70)	2,800
(40 × 8)	320
(3 × 70)	210
(3 × 8)	24
	3,354

1. 29
× 14

2. 34
× 51

3. 62
× 22

4. 44
× 36

5. 81
× 53

6. 39
× 28

7. 76
× 64

Use estimation to solve each problem. Write a number model to show how you estimate.

8. Dina has $25.00. Does she have enough to buy a radio that costs $14.79 and a CD that costs $11.25?

_____

9. Jaime has $40.00. Does he have enough to buy shoes that cost $21.97 and two ties that cost $8.50 each?

_____

10. Janice has $50.00. How many plants can she buy if each plant costs $11.95?

_____

11. Jack has $60.00. How many shirts can he buy if each shirt costs $14.50?

_____

12. **Writing/Reasoning** Jason has $63.00 in his savings account. He received $35.00 as a birthday gift. Does he have enough money to buy a bike that costs $90.00? Explain your answer.

_____

Practice Set 82 continued

SRB
31 32
250–253

Solve each problem.

13. Arthur bought a goldfish for 49¢, a striped fish for $0.72, and fish food for 68¢. How much did Arthur spend?

_____

14. How much change did Arthur receive if he paid for the 3 items with $3.00?

_____

15. Betty wants to buy a dog collar for $3.50, a water dish for $2.79, and a toy bone for $3.49. Can she buy all 3 items with $10.00?

_____

16. How much do the 3 items that Betty wants to buy cost altogether?

_____

Write <, >, or =.

17. This is ONE:

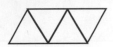

$\frac{2}{4}$ —— $\frac{1}{4}$ $\frac{3}{4}$ —— $\frac{1}{2}$

18. This is ONE:

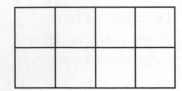

$\frac{5}{8}$ —— $\frac{1}{2}$ $\frac{1}{4}$ —— $\frac{2}{8}$

19. This is ONE:

$\frac{2}{3}$ —— $\frac{4}{6}$ $\frac{1}{3}$ —— $\frac{1}{2}$

Use the partial-products algorithm to solve the following problems.

20. 378
      $\times$ 6

21.   60
      $\times$ 45

22. 50 $\times$ 45 = _____

Use with or after Lesson 9·12.

Practice Set 83

SRB
11
171–173

Write the number and the unit for each problem.
Use Celsius temperatures.

Example   4 degrees below zero **−4°C**

1. 25 degrees above zero _____

2. 58 degrees below zero _____

3. zero degrees _____

4. 150 degrees above zero _____

5. 14 degrees below zero _____

6. 100 degrees below zero _____

Which temperature is colder? You can use the thermometer to help you.

Example   −6°C or −14°C
            −14°C is below −6°C on the thermometer.
            −14°C is colder than −6°C.

7. 0°C or 5°C

8. 2°C or −20°C

_____   _____

9. 9°C or −9°C

10. −7°C or 0°C

_____   _____

Which temperature is warmer? You can use the thermometer to help you.

Example   24°C or −2°C
            24°C is above −2°C on the thermometer.
            24°C is warmer than −2°C.

11. 0°C or −8°C

12. −98°C or 1°C

_____   _____

13. 15°C or −15°C

14. 12°C or −35°C

_____   _____

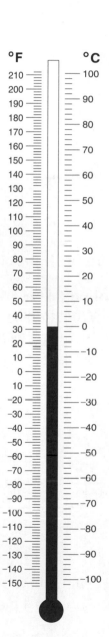

Practice Set 83 continued

Complete the list of factors for each number below.

> *The factors of a number* are the numbers that
> can be multiplied by whole numbers to get that
> number, or the numbers that a number can be
> divided by without having remainders.

Example Factors of 16: 1, __2__, __4__, 8, __16__

15. Factors of 7: _____, 7

16. Factors of 18: _____, 2, _____, _____, _____, 18

17. Factors of 36: 1, _____, _____, 4, _____, 9, _____, _____, 36

18. Factors of 50: _____, 2, _____, _____, 25, _____

Solve each problem.

19. Jerry went swimming 19 days in June. What fraction of the days in June did Jerry go swimming? (*Hint:* June has 30 days.)

20. Sarah spent 20 minutes eating breakfast. What fraction of an hour did she spend eating breakfast? What fraction of an hour did she NOT spend eating breakfast?

_____

_____

21. **Writing/Reasoning** Explain or show how you found your answers to Problem 20.

22. Sam and two friends shared a pizza cut into 8 pieces. Sam ate 1 piece, and each of his friends ate 2 pieces.

What fraction of the pizza did Sam eat? _____

What fraction of the pizza did each friend eat? _____

What fraction of the pizza was left over? _____

Use with or after Lesson 9·13.

Practice Set 84

Find each missing number. Use fractions.

1 meter = 10 decimeters	1 yard = 3 feet
1 meter = 100 centimeters	1 yard = 36 inches
1 decimeter = 10 centimeters	1 foot = 12 inches
1 centimeter = 10 millimeters	

1. _____ yard = 12 inches

2. _____ meter = 8 decimeters

3. 1 foot = _____ yard

4. 50 centimeters = _____ meter

5. 9 inches = _____ foot

6. _____ centimeter = 3 millimeters

7. _____ yard = 2 feet

8. 9 centimeters = _____ meter

9. **Writing/Reasoning** Give three different names for 18 inches. Explain how you found each of your answers.

_____

_____

_____

_____

Find each answer. You can draw pictures or use counters.

10. There are 12 students taking swimming lessons. $\frac{1}{3}$ of them are third graders. How many are third graders?

_____

11. The pet store has 10 dogs for sale. Half of the dogs are collies. How many of the dogs are collies?

_____

12. Karen drew a picture of 8 flags. She colored $\frac{1}{4}$ of the flags orange.

How many flags did she color orange? _____

What fraction of the flags did she NOT color orange? _____

Practice Set 84 *continued*

For each ruler, find the distance between the two points.

Example   *A* to *B*
The distance between *A* and *B* is $2\frac{1}{4}$ in.

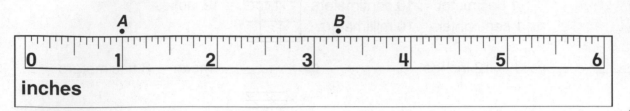

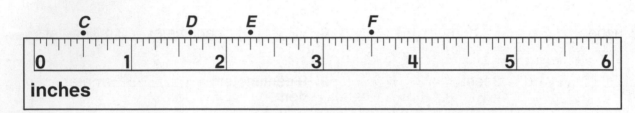

13. *C* to *D*

_____

14. *E* to *F*

_____

15. *C* to *F*

_____

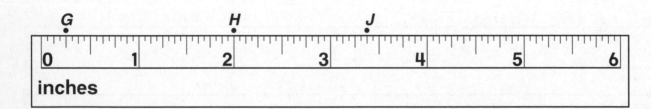

16. *G* to *H*

_____

17. *H* to *J*

_____

18. *G* to *J*

_____

 Multiply. Remember to practice and memorize your multiplication facts.

19. $4 \times 1 =$ _____

20. $3 \times 9 =$ _____

21. $9 \times 10 =$ _____

22. $5 \times 2 =$ _____

23. $10 \times 10 =$ _____

24. $4 \times 8 =$ _____

25. $7 \times 9 =$ _____

26. $10 \times 3 =$ _____

27. $6 \times 3 =$ _____

28. $9 \times 5 =$ _____

29. $4 \times 2 =$ _____

30. $8 \times 6 =$ _____

Use with or after Lesson 10·1.

Practice Set 85

Find the volume (V) of each box. Each cube stands for 1 cubic centimeter.

1.   V = _____

2.   V = _____

3. 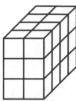   V = _____

4. 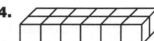   V = _____

Tell which unit you would use to measure each item. Choose from inch, foot, yard, and mile.

> **Example**   the length of a pencil
> **Unit: inch**

5. the length of your math book _____

6. the length of a paper clip _____

7. the distance between Chicago and St. Louis _____

8. the length of a football field _____

9. the width of your hand _____

10. the width of a room _____

11. the width of your foot _____

12. the width of a park _____

13. the distance traveled in a car after one hour _____

14. the height of a dog _____

Practice Set 85 *continued*

SRB
11
171–173

Which temperature is colder? You can use the thermometer to help you.

15. 0°F or −18°F _____

16. 12°F or −12°F _____

17. 0°F or 6°F _____

18. −8°F or −18°F _____

19. 5°F or 15°F _____

20. −23°F or −32°F _____

Which temperature is warmer? You can use the thermometer to help you.

21. 6°C or 36°C _____

22. −14°C or −45°C _____

23. 0°C or −10° C _____

24. 12°C or −12°C _____

25. 16°C or −37°C _____

26. 20°C or 0°C _____

Solve each problem. You can use the thermometer to help you.

27. One January morning, the temperature was −18°F. By noon, the temperature had risen to 4°F. How many degrees had the temperature risen?

_____

28. One June morning, the temperature was 18°C. By 2:00 in the afternoon, the temperature had risen to 34°C. How many degrees had the temperature risen?

_____

29. On Tuesday, the high temperature was −10°C. On Friday, the high temperature was 4°C. How many degrees warmer was the high temperature on Friday?

_____

Use with or after Lesson 10·2.

Practice Set 86

SRB
162–166

Read the scale and record the weight.

1.

_____

2.

_____

3.

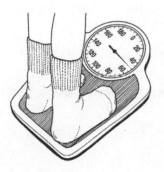

_____

4.

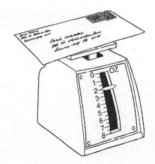

_____

Find the volume of the rectangular prisms.

5.

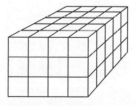

_____ cubic centimeters

6.

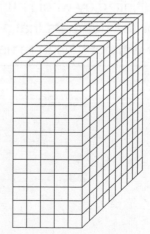

_____ cubic centimeters

Practice Set 86 continued

SRB
37–40

Find the missing numbers. Use fractions.

> 1 meter = 10 decimeters   1 yard = 3 feet
> 1 meter = 100 centimeters   1 yard = 36 inches
> 1 decimeter = 10 centimeters   1 foot = 12 inches
> 1 centimeter = 10 millimeters

Example   3 centimeters = $\dfrac{3}{100}$ meter

7. 1 inch = _____ foot

8. _____ meter = 1 centimeter

9. 1 foot = _____ yard

10. _____ decimeter = 1 centimeter

11. _____ meter = 1 decimeter

12. 6 centimeters = _____ meter

13. 12 inches = _____ foot

14. _____ centimeter = 1 millimeter

15. 2 feet = _____ yard

16. 70 centimeters = _____ meter

17. _____ yard = 24 inches

18. 9 decimeters = _____ meter

Find the factors for each number listed below.

> *The factors of a number* are the numbers that
> can be multiplied by whole numbers to get that
> number, or the numbers that a number can be
> divided by without having remainders.

19. 9

20. 20

21. 35

_____

_____

_____

22. 60

23. 8

24. 51

_____

_____

_____

25. 32

26. 45

27. 21

_____

_____

_____

Use with or after Lesson 10·4.

Practice Set 87

SRB 160 161

**Tell which unit you would use to measure each item.
Choose from gallon, quart, pint, cup, ounce, and tablespoon.**

1. amount of water you drink with dinner _____

2. container of milk that you buy at the store_____

3. amount of water in a bathtub _____

4. amount of juice in a can from a vending machine _____

5. amount of syrup on pancakes _____

6. container of orange juice that you buy at the store _____

7. amount of water in an eyedropper _____

8. amount of water in a swimming pool _____

9. amount of cream in a cup of coffee _____

10. amount of lemonade needed to serve 4 people _____

FACTS PRACTICE

**Multiply. Remember to practice and memorize
your multiplication facts.**

11. $9 \times 9 =$ _____

12. $8 \times 7 =$ _____

13. $6 \times 8 =$ _____

14. $9 \times 6 =$ _____

15. $7 \times 7 =$ _____

16. $8 \times 9 =$ _____

17. $7 \times 6 =$ _____

18. $8 \times 8 =$ _____

19. $7 \times 9 =$ _____

20. 400 [800s] = _____

21. 100 [700s] = _____

22. 200 [500s] = _____

23. 300 [400s] = _____

24. 500 [200s] = _____

25. 100 [900s] = _____

26. $40 \times 70 =$ _____

27. $60 \times 80 =$ _____

28. $90 \times 90 =$ _____

29. $700 \times 600 =$ _____

30. $800 \times 900 =$ _____

31. $500 \times 600 =$ _____

Practice Set 88

SRB
83–85
146 147

Find the mean for each data set below.

Example	9 5 7 6 3
Step 1	Find the total of the numbers in the data set.
	9 + 5 + 7 + 6 + 3 = 30
Step 2	Count the numbers in the data set.
	There are 5 numbers in all.
Step 3	Divide the total by 5.
	30 ÷ 5 = 6
	The mean is 6.

1. 7 2 5 6

_____

2. 5 4 2 5 6 2

_____

3. 12 8 7 10 13

_____

4. 9 5 6 10 10 11 12

_____

Find the median for each data set below.

Example   58, 63, 65, 49, 58, 73, 61, 65

To find the **median,** put the numbers in order from least to greatest. The median is the number with an equal number of values above and below it.

49, 58, 58, 61, 63, 65, 65, 73
The median is between 61 and 63.

5. 7, 2, 5, 6

_____

6. 5, 4, 2, 5, 6, 2, 7

_____

Find each missing number.

| 1 mile (mi) = 1,760 yards (yd) |
| 1 mile (mi) = 5,280 feet (ft) |
| 1 yard (yd) = 3 feet (ft) |
| 1 yard (yd) = 36 inches (in.) |
| 1 foot (ft) = 12 inches (in.) |

7. _____ feet = 2 yards

8. 18 inches = _____ ft _____ in.

9. 3 yd 2 ft = _____ ft

10. 9 ft = _____ yd _____ in.

11. 2 miles = _____ yards

12. 10,560 feet = _____ miles

13. 75 in. = _____ yd _____ in.

14. 3 ft 9 in. = _____ in.

Use with or after Lesson 10·7.

Practice Set 88 continued

Use the bar graph to answer each question below.

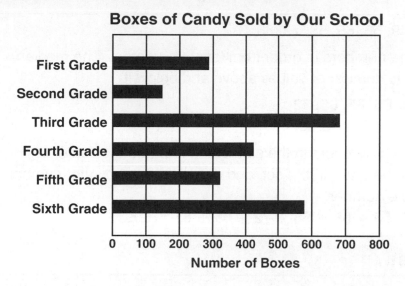

Boxes of Candy Sold by Our School

15. Which grade sold the fewest boxes of candy?

_____

About how many boxes of candy did they sell?

_____

16. Which grade sold the most boxes of candy?

_____

About how many boxes of candy did they sell?

_____

17. Estimate the range, or difference between the highest and lowest numbers, on the graph.

_____

18. Which grade sold a little more than twice as many boxes of candy as the second grade sold?

_____

19. Put the grades in order from the grade that sold the most boxes to the grade that sold the fewest boxes.

_____

20. About how many more boxes of candy would the fourth grade have to sell in order to reach 500 boxes?

_____

Practice Set 89

Find the median and the mean for each data set below.

Example   58, 63, 65, 49, 53, 65, 58, 72, 65, 61

To find the **median,** put the numbers in order from least to greatest. The median is the number with an equal number of values above and below it.

    49, 53, 58, 58, 61, 63, 65, 65, 65, 72
    The median is between 61 and 63.

To find the **mean,** add all the numbers in the data set. (*Hint:* Use a calculator.)
Then count how many numbers are in the set, and divide the total by that number.
Round to the nearest whole number.

    58 + 63 + 65 + 49 + 53 + 65 + 58 + 72 + 65 + 61 = 609
    609 ÷ 10 = 60.9
    The mean is about 61.

---

1. 195, 204, 198, 187, 200, 193, 205, 187, 182

2. 929, 905, 917, 933, 904, 913, 905, 901, 908, 922

3. 397, 395, 400, 403, 397, 382, 410, 406, 395, 397

4. 1,111; 1,083; 1,102; 1,075; 1,096; 1,114; 1,111; 1,075

5. 15,270; 15,400; 15,230; 15,320; 15,290; 15,405; 15,300; 15,240; 15,320

144

Practice Set 89 *continued*

Match each measurement in the first list with an equal measurement in the second list. Write the letter that identifies that equal measurement.

1 gallon	=	4 quarts
1 quart	=	2 pints
1 pint	=	2 cups
1 cup	=	8 fluid ounces

6. 2 gallons _____   **A.** 4 pints

7. 5 cups _____   **B.** $\frac{1}{16}$ gallon

8. 4 fluid ounces _____   **C.** 24 fluid ounces

9. $2\frac{1}{2}$ pints _____   **D.** 6 pints

10. $\frac{1}{2}$ gallon _____   **E.** 5 cups

11. 3 pints _____   **F.** $\frac{1}{2}$ cup

12. 3 quarts _____   **G.** 6 quarts

13. $1\frac{1}{2}$ gallons _____   **H.** 8 quarts

14. 3 cups _____   **I.** 40 fluid ounces

15. 1 cup _____   **J.** 6 cups

Write the value of the underlined digit in each number.

16. 479,2̲14   **17.** 289.4̲6̲   **18.** 7̲8,432

_____   _____   _____

19. 3̲.289   **20.** 1̲0̲8.27   **21.** 1̲29,568

_____   _____   _____

FACTS PRACTICE

Multiply. Remember to practice and memorize your multiplication facts.

22. $2 \times 3 =$ _____   **23.** $8 \times 3 =$ _____   **24.** $7 \times 5 =$ _____   **25.** $9 \times 9 =$ _____

26. $7 \times 10 =$ _____   **27.** $1 \times 1 =$ _____   **28.** $4 \times 6 =$ _____   **29.** $8 \times 9 =$ _____

30. $6 \times 10 =$ _____   **31.** $2 \times 7 =$ _____   **32.** $3 \times 3 =$ _____   **33.** $8 \times 1 =$ _____

Practice Set 90

The frequency table below shows the number of school lunches bought in one week by students in different classrooms. Use a calculator to help you answer each question.

Room	Number of School Lunches Bought
101	~~HHH~~ ~~HHH~~ ~~HHH~~
102	~~HHH~~ ~~HHH~~
103	~~HHH~~ ///
104	~~HHH~~ //
105	~~HHH~~ ~~HHH~~ /
106	~~HHH~~
107	~~HHH~~ ~~HHH~~
108	~~HHH~~ /

1. What is the total number of school lunches bought?

_____

2. Find the **mode,** or the number that occurs most often, in the data.

_____

3. Find the **median** number of school lunches bought.

_____

4. Find the **mean** number of school lunches bought.

_____

5. **Writing/Reasoning** Explain how you found the **mean** in Problem 4.

_____

_____

_____

Solve each problem.

6. Jason had 25 quarters. He put 7 of them in his bank. What fraction of the quarters did Jason put in his bank?

_____

7. Becky spends 5 hours each day at school. What fraction of the day does Becky spend at school? (*Hint:* A day has 24 hours.)

_____

8. Bryan has 4 history videos, 5 science videos, and 3 adventure videos. What fraction of his videos are history? What fraction of his videos are adventure?

_____ are history.   _____ are adventure.

Practice Set 90 ◆ *continued*

Draw and shade shapes to show each fraction.

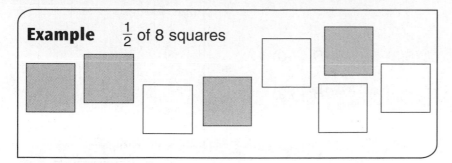

Example $\frac{1}{2}$ of 8 squares

9. $\frac{1}{4}$ of 4 triangles

10. $\frac{1}{3}$ of 6 circles

11. $\frac{3}{4}$ of 8 rectangles

12. $\frac{1}{2}$ of 10 diamonds

13. $\frac{2}{3}$ of 6 triangles

14. $\frac{1}{4}$ of 8 circles

Solve each problem.

15. Felix is following a recipe that calls for 3 cups of milk. How many cups of milk does he need to double the recipe?

_____

16. Sharon bought 3 gallons of juice. There are 4 quarts in 1 gallon. How many quarts of juice did Sharon buy?

_____

17. At the end of the school year, Lisa weighed 62 pounds. She had gained 6 pounds during the school year. How much did Lisa weigh at the beginning of school?

_____

18. Sarah rode her bike 18 kilometers Monday and 25 kilometers Tuesday. How many kilometers did Sarah ride her bike in those two days?

_____

Practice Set 91

Follow the directions.

1. Plot these points:

   A: (1,1)   B: (3,6)   C: (6,3)   D: (7,0)

   E: (5,6)   F: (7,9)   G: (9,6)   H: (3,9)

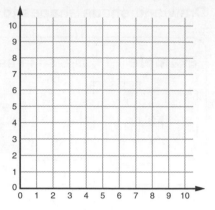

2. Draw the following line segments: $\overline{AB}, \overline{BC}, \overline{CA}$

   What shape did you make? _____

3. Draw the following line segments: $\overline{DE}, \overline{EF}, \overline{FG}, \overline{GD}$

   What shape did you make? _____

4. Measure the following line segments to the nearest $\frac{1}{2}$ centimeter: $\overline{AB}, \overline{DG}, \overline{FG}, \overline{CE}$

   _____

Find the volume (V) of each box. Each cube stands for 1 cubic inch.

5.   V = _____

6.   V = _____

7.    V = _____

8. 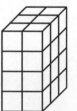   V = _____

 Multiply. Remember to practice and memorize your multiplication facts.

9. $8 \times 10 =$ _____   10. $9 \times 4 =$ _____   11. $7 \times 3 =$ _____   12. $2 \times 2 =$ _____

13. $1 \times 5 =$ _____   14. $7 \times 7 =$ _____   15. $6 \times 5 =$ _____   16. $10 \times 7 =$ _____

Use with or after Lesson 10·11.

Practice Set 92

Five students recorded their heights. The data are below.

Student Names and Heights				
Jose 49 in.	Thomas 52 in.	Binta 51 in.	Mario 57 in.	Amy 53 in.

1. Use the information in the table to complete the bar graph.

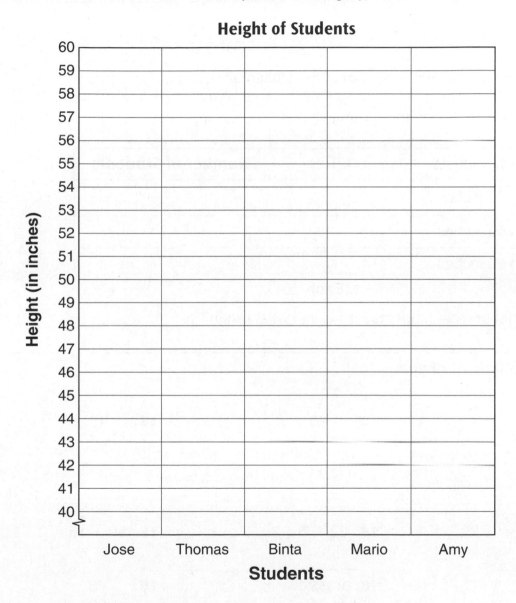

Look at your graph to answer Problems 2 and 3.

2. Which student is the shortest? _____ What is the height of the student? _____

3. Which student is the tallest? _____ What is the height of the student? _____

Practice Set 92 continued

Use the graph on page 149 to answer Problems 4–9.

4. What is the maximum height of the students? _____

5. What is the minimum height of the students? _____

6. What is the mean (average) height of the students? _____

7. What is the median height of the students? _____

8. What is the range of the heights of the students? _____

9. Use the information in the graph to make a frequency table of the students' heights.

Height Ranges	Tallies	Number of Students
46 to 50 inches		
51 to 55 inches		
56 to 60 inches		

10. The height of most students falls within what range? _____

Multiply or divide.

11.   34
   × 22

12.   60
   × 15

13.   18
   × 5

14. 75 ÷ 5 = _____

15. 32 ÷ 8 = _____

16. 90 ÷ 15 = _____

17.   12
   × 18

18.   53
   × 31

19.   48
   × 20

20. 81 ÷ 9 = _____

21. 63 ÷ 7 = _____

22. 72 ÷ 8 = _____

Use with or after Lesson 11·2.

Practice Set 93

Imagine that each of the following containers is tipped over onto a table.

1. How many *heads* do you expect to fall face up?

_____

How many *tails* do you expect to fall face up?

_____

2. How many *black* sides do you expect to fall face up?

_____

How many *white* sides do you expect to fall face up?

_____

3. How many *front* sides do you expect to fall face up?

_____

How many *back* sides do you expect to fall face up?

_____

**Solve each division problem. If the problem has a remainder,
write that amount after the letter *R*.**

Example $100 \div 9 \rightarrow$ **11 R1**

4. $81 \div 9 \rightarrow$ _____

5. $54 \div 6 \rightarrow$ _____

6. $72 \div 8 \rightarrow$ _____

7. $56 \div 8 \rightarrow$ _____

8. $48 \div 6 \rightarrow$ _____

9. $36 \div 6 \rightarrow$ _____

10. $35 \div 4 \rightarrow$ _____

11. $200 \div 6 \rightarrow$ _____

12. $95 \div 4 \rightarrow$ _____

13. $17 \div 6 \rightarrow$ _____

14. $3,000 \div 5 \rightarrow$ _____

15. $332 \div 10 \rightarrow$ _____

16. $4,203 \div 3 \rightarrow$ _____

17. $73 \div 9 \rightarrow$ _____

18. $721 \div 10 \rightarrow$ _____

Practice Set 93 *continued*

19. Use only blue, red, and green. Make a spinner so that
the paper clip

- has the same chance of landing on each color.

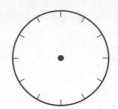

20. Use only red, yellow, and blue. Make a spinner so that
the paper clip

- has the greatest chance of landing on red

and

- has a lesser chance of landing on blue

and

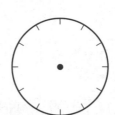

- has the least chance of landing on yellow.

21. The Jackson family left home at the time
shown on the clock on the left. They arrived
at their destination at the time shown on the
clock on the right. How long did their trip take?

_____

22. It takes Tara 15 minutes to walk to the library
from her home. She spends 45 minutes at the
library. Tara walks back home. How long was
she away from home?

_____

The shaded part of each clock shows passing time.

23.

How many minutes have passed?

_____

24.

How many minutes have passed?

_____

152

Practice Set 94

SRB
83–85
92–94

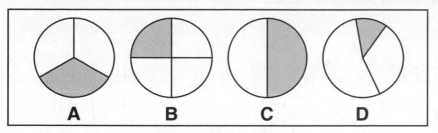

A   B   C   D

On which of the spinners above …

1. are you equally likely to land on the shaded part or the white part?

_____

2. are you likely to land on the shaded part about $\frac{1}{4}$ of the time?

_____

3. are you twice as likely to land on a white part?

_____

4. are you likely to land on a white part about $\frac{3}{4}$ of the time?

_____

5. are you likely to land on the shaded part about $\frac{1}{3}$ of the time?

_____

Find the mean of each data set below.

> To *find the mean*, add all the numbers in the data set. Then count how many numbers are in the set, and divide the total by that number. Round to the nearest whole number.

6. 349, 756, 821, 444, 348, 259 _____

7. 3,500; 3,511; 3,487; 3,548 _____

8. 28, 34, 56, 54, 76, 89, 21, 13, 49, 112 _____

9. 1,001; 1,012; 998; 799; 804; 1,030 _____

Practice Set 94 continued

Write the missing numbers.

10.

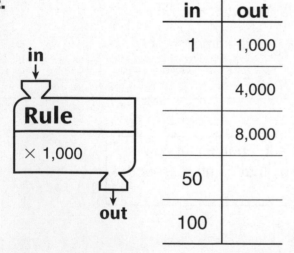

in	out
1	1,000
2	
	3,000
7	
	10,000

11.

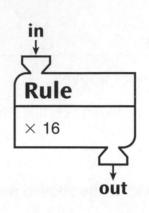

in	out
1	16
	32
	64
7	
	160

12.

in → Rule × 1,000 → out

in	out
1	1,000
	4,000
	8,000
50	
100	

13.

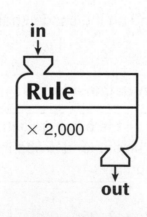

in	out
1	2,000
2	
4	
	20,000
7	

Use with or after Lesson 11·4.

Practice Set 95

Make the following predictions:

1. You make 10 random draws. *You draw …*
   • blue 4 times
   • red 4 times
   • yellow 2 times

   Predict the colors of the 5 blocks in the bag at the right. _____

   Tell what fraction of the blocks is NOT yellow. _____

2. You make 25 random draws. *You draw …*
   • orange 16 times
   • purple 9 times

   Predict the colors of the 5 blocks in the bag. _____

3. You make 50 random draws. *You draw …*
   • red 19 times
   • blue 21 times
   • white 10 times

   Predict the colors of the 5 blocks in the bag. _____

   Tell what fraction of the blocks are white. _____

4. You make 50 random draws. *You draw …*
   • blue 26 times
   • red 24 times

   Predict the colors of the 6 blocks in the bag at the right.

   _____

5. You make 40 random draws. *You draw …*
   • pink 6 times
   • orange 13 times
   • red 7 times
   • green 14 times

   Predict the colors of the 6 blocks in the bag. _____

   Tell what fraction of the blocks are pink. _____

Practice Set 95 continued

**Add numbers to the chart and find the total for each column.
Then answer the questions.**

There are 5 third-grade classes at Lincoln Elementary.
- Room 101 has 12 boys and 12 girls.
- Room 102 has 14 boys and 12 girls.
- Room 103 has 13 boys and 13 girls.
- Room 104 has 11 boys and 13 girls.
- Room 105 has 12 boys and 11 girls.

6.

Number of Third Graders at Lincoln Elementary School		
Room Number	Boys	Girls
101		
102		
103		
104		
105		
TOTALS		

7. Use fractions to tell about how many of the third-grade students are girls and about how many are boys.

_____

8. There are about 100 *second* graders at the school. Predict how many are boys and how many are girls.

_____

9. There are about 130 *fourth* graders at the school. Predict how many are girls and how many are boys.

_____

10. Can you predict whether a new student in the third grade will be a boy or a girl?

_____

11. **Writing/Reasoning** Explain your answer to Problem 10.

_____

_____

_____

Test Practice ⟨ 1 ⟩

Fill in the circle next to your answer.

1. This chart shows the amounts raised by children in each grade for a school fundraiser. Which of the following correctly compares the amounts of money raised by Grade 3 and Grade 5?

   Ⓐ $1,264 = $1,259

   Ⓑ $1,264 + $1,259

   Ⓒ $1,264 > $1,259

   Ⓓ $1,264 < $1,259

Amounts Raised	
Grade	**Amount (in dollars)**
2	$1,248
3	$1,264
4	$1,285
5	$1,259

2. What are the next 3 dates in this pattern?

   Ⓐ September 15, 18, and 21

   Ⓑ September 14, 16, and 18

   Ⓒ September 13, 14, and 15

   Ⓓ September 16, 20, and 24

3. Robin is sewing fabric squares together to make this quilt. Each square is 1 foot by 1 foot.

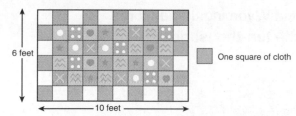

   6 feet

   10 feet

   ☐ One square of cloth

   How many fabric squares will Robin need to make the quilt?

   Ⓐ 10   Ⓑ 16   Ⓒ 32   Ⓓ 60

4. Beth is writing the fact family for the numbers 6, 7, and 13 on her Fact Triangle.

   $$13 - 7 = 6 \qquad 6 + 7 = 13 \qquad 13 - 6 = 7$$

   Which fact is missing from her list?

   Ⓐ $13 - 8 = 5$ Ⓑ $7 + 6 = 13$ Ⓒ $7 - 6 = 1$ Ⓓ $6 + 7 = 42$

5. Look at this pattern of dates. What date is missing?

   May 4, 10, 16, _____, 28

   Ⓐ 14   Ⓑ 18   Ⓒ 22   Ⓓ 26

Test Practice 1 *continued*

Fill in the circle next to your answer.

6. At 8:00 A.M. the temperature in Key West was 63°F. The
temperature in Pensacola was 41°F. Which of the following
correctly shows the difference between the two temperatures?

  Ⓐ $8 + 63 = 8 + 41$ Ⓑ $41 + 8 < 63$

  Ⓒ $63 > 41 + 22$ Ⓓ $63 - 41 = 22$

7. Mr. Brinkman's class will arrive at the
museum two hours after it opens.
This clock shows the time the museum
opens. What time will they arrive at the
museum?

  Ⓐ 7:00 A.M.   Ⓑ 9:00 A.M.   Ⓒ 11:00 A.M.   Ⓓ 1:00 A.M.

8. Gannett Peak is the highest point in Wyoming. It measures
13,804 feet tall. Which digit is in the **ten-thousands**
place in 13,804?

  Ⓐ 1   Ⓑ 3   Ⓒ 4   Ⓓ 8

9. The picture below shows a rake next to a sidewalk.
The rake is 5 feet long. About how long is the sidewalk?

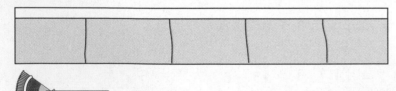

  Ⓐ 10 feet   Ⓑ 20 feet   Ⓒ 30 feet   Ⓓ 40 feet

Test Practice 2

Fill in the circle next to your answer.

1. Which situation matches the equation below?

   $$9 \times 3 = 27$$

   Ⓐ Alice had 27 photos. She placed 3 photos in her album. How many photos have not been placed in the album?

   Ⓑ Alice placed 3 photos in her album. Then she placed 9 more in the album. How many photos are in Alice's album now?

   Ⓒ Alice had 27 photos. She placed 9 photos on each page of her album. How many pages did Alice fill?

   Ⓓ Alice filled 9 pages in her album. She placed 3 photos on each page. How many photos does she have in all?

2. This picture shows that 3 and 4 are factors of 12.

   4 ⋮⋮⋮
   3

   Which of the following shows factors of 27?

   Ⓐ 2 •••••••••••••
   13

   Ⓑ 4 ••••••••••
   8

   Ⓒ 4 •••••••
   7

   Ⓓ 3 •••••••••
   9

3. This Fact Triangle shows a fact family. Which of the following is NOT a member of this fact family?

   Ⓐ 42 ÷ 6 = 7

   Ⓑ 3 × 2 = 6

   Ⓒ 42 ÷ 7 = 6

   Ⓓ 7 × 6 = 42

   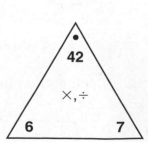

Test Practice 2 continued

Fill in the circle next to your answer.

4. Which streets are parallel?

Ⓐ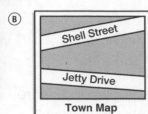
Town Map

Ⓑ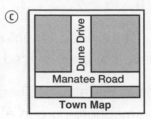
Town Map

Ⓒ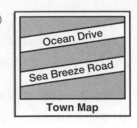
Town Map

Ⓓ
Town Map

5. Which amount is equivalent to 3 dollar bills, 5 dimes, and 2 pennies?

Ⓐ $3.51   Ⓑ $3.52   Ⓒ $3.53   Ⓓ $3.25

6. Which of these stained glass windows has a picture that is **symmetric?**

Ⓐ
Ⓑ
Ⓒ
Ⓓ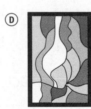

7. Which shape is made up of 27 cubes?

Ⓐ
Ⓑ
Ⓒ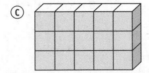
Ⓓ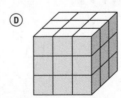

8. Which figure has right angles?

Ⓐ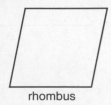
rhombus

Ⓑ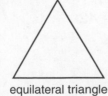
equilateral triangle

Ⓒ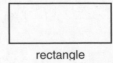
rectangle

Ⓓ
octagon

160

Test Practice 3

Fill in the circle next to your answer.

1. Nick is buying these items at the science shop. Which of the following is the best estimate of his total cost?

   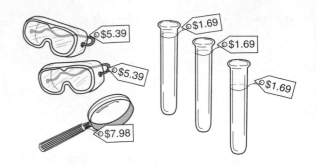

   Ⓐ Less than $22

   Ⓑ Between $23 and $26

   Ⓒ Between $27 and $31

   Ⓓ More than $32

2. Mr. Williams charges $40 to mow each customer's lawn. How much money will Mr. Williams earn if he has 8 customers?

   Ⓐ $5   Ⓑ $48   Ⓒ $320   Ⓓ $408

3. Matt leaves school and stops at the park on his way home. The park is $\frac{4}{6}$ of the way to his home.

   Which fraction is equal to $\frac{4}{6}$?

   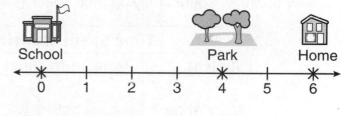

   Ⓐ $\frac{1}{6}$ Ⓑ $\frac{1}{3}$ Ⓒ $\frac{1}{2}$ Ⓓ $\frac{2}{3}$

4. Grace fills each magazine holder with 20 magazines. Then she places the magazine holder on the shelf. About how many magazines will fill the shelf?

   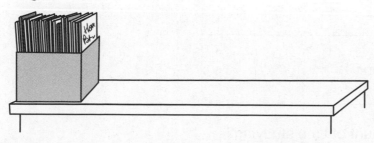

   Ⓐ 20   Ⓑ 50

   Ⓒ 80   Ⓓ 120

Test Practice 3 continued

Fill in the circle next to your answer.

5. This much pizza was left over after Salvador ate.

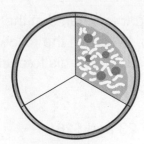

Emily's pizza is sliced as shown. If Emily eats
the same amount of pizza as Salvador,
how many pieces will be left over?

(A) 1   (B) 2

(C) 3   (D) 4

6. Mr. Montales made the chart below. It shows how much time
four students spent studying their spelling words.

Time Spent Studying	
Student	**Time (in fractions of an hour)**
Mary Anne	$\frac{1}{3}$ hour
Jared	$\frac{1}{2}$ hour
Erin	$\frac{1}{8}$ hour
Bill	$\frac{1}{5}$ hour

Which student spent the **least** amount of time studying?

(A) Mary Anne   (B) Jared

(C) Erin   (D) Bill

162

Use with or after Unit 9.

Test Practice  3 continued

7. Each of these shapes has a perimeter of 24 centimeters. Which shape has the **greatest** area?

Ⓐ

Ⓑ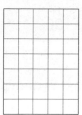

Ⓒ

Ⓓ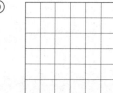

8. Last week, 46 third graders signed up for summer camp. Each cabin holds 6 children. How many cabins are needed?

Ⓐ 7 Ⓑ 8 Ⓒ 40 Ⓓ 52

9. Ed, Kay, Max, and Jasmine sold the pencils shown below at the school fair. They each sold an equal number of pencils. How many pencils were left over?

Pencils for Sale

Ⓐ 1 Ⓑ 2 Ⓒ 3 Ⓓ 4

Test Practice ❰ 4 ❱

Fill in the circle next to your answer.

1. How many hours are in 2 days?

  Ⓐ 12   Ⓑ 24   Ⓒ 36   Ⓓ 48

2. These are the ages of the 6 art contest winners.

            8, 6, 5, 7, 6, 10

What is the **range** of the ages?

  Ⓐ 5   Ⓑ 6

  Ⓒ 7   Ⓓ 10

3. Donna rolled 2 number cubes and wrote down the sum.
She rolled 5 times. Look at Donna's data.

           Sum: 7, 5, 2, 7, 10

Which statement is true about this set of data?

  Ⓐ mode = median   Ⓑ median > mode

  Ⓒ mode < median   Ⓓ median < mode

4. There are 10 chance cards in a game.

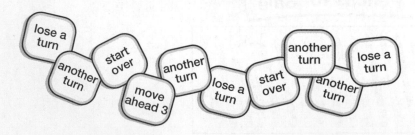

Ross shuffles the cards and places them facedown.
Which card is **most** likely to be picked?

  Ⓐ move ahead 3   Ⓑ start over

  Ⓒ another turn   Ⓓ lose a turn

Test Practice ◆ 4 ▷ *continued*

Fill in the circle next to your answer.

5. Jorge made a graph to show some of the children's favorite sports at Crest Ridge School. Which of the following is true?

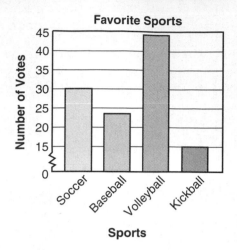

Favorite Sports

    Ⓐ Soccer is the most popular sport.

    Ⓑ Volleyball received more than 50 votes.

    Ⓒ Kickball received the fewest number of votes.

    Ⓓ Baseball is twice as popular as kickball.

6. Stephen runs around the track each day at school. After school, he writes down the number of laps he ran.

MONDAY	TUESDAY	WEDNESDAY	THURSDAY	FRIDAY
4	3	6	7	5

What is the **mean** of these numbers?

    Ⓐ 3   Ⓑ 5   Ⓒ 7   Ⓓ 25

7. Mrs. Martinez asked the children in her class which instrument was their favorite. A total of 10 children chose the flute.

Favorite Instruments

Instrument	Number of Children
Oboe	♪♪♪
Flute	
Clarinet	♪♪

Key
♪ = 2 votes

In this pictograph, how many notes should there be for the flute?

    Ⓐ 2   Ⓑ 3   Ⓒ 5   Ⓓ 10